NOTES ON THE
BINOMIAL TRANSFORM

Theory and Table with Appendix on Stirling Transform

NOTES ON THE BINOMIAL TRANSFORM

Theory and Table with Appendix on Stirling Transform

Khristo N Boyadzhiev

Ohio Northern University, USA

World Scientific

NEW JERSEY · LONDON · SINGAPORE · BEIJING · SHANGHAI · HONG KONG · TAIPEI · CHENNAI · TOKYO

Published by

World Scientific Publishing Co. Pte. Ltd.

5 Toh Tuck Link, Singapore 596224

USA office: 27 Warren Street, Suite 401-402, Hackensack, NJ 07601

UK office: 57 Shelton Street, Covent Garden, London WC2H 9HE

Library of Congress Cataloging-in-Publication Data
Names: Boyadzhiev, Khristo N., author.
Title: Notes on the binomial transform : theory, and table with appendix on Stirling transform /
    by Khristo N. Boyadzhiev (Ohio Northern University, USA).
Description: New Jersey : World Scientific, 2018. | Includes bibliographical references and index.
Identifiers: LCCN 2018012480 | ISBN 9789813234970 (hardcover : alk. paper)
Subjects: LCSH: Transformations (Mathematics) | Combinatorial analysis.
Classification: LCC QA613.7 .B69 2018 | DDC 511/.6--dc23
LC record available at https://lccn.loc.gov/2018012480

British Library Cataloguing-in-Publication Data
A catalogue record for this book is available from the British Library.

For any available supplementary material, please visit
http://www.worldscientific.com/worldscibooks/10.1142/10848#t=suppl

Printed in Singapore

This book is dedicated to my wife Irina and daughters Marinella and Alexandra.

Preface

The binomial transform is a discrete transformation of one sequence into another with many interesting applications in combinatorics and analysis. The binomial transform theory belongs to the expanding field of discrete mathematics. The material of this book will be helpful to researchers interested in enumerative combinatorics, special numbers, and classical analysis.

The book can also serve as a valuable reference. It can be used as lecture notes for a course on binomial identities, binomial transforms and Euler series transformations. The binomial transform leads to various combinatorial and analytical identities involving binomial coefficients. Many interesting identities can be written as binomial transforms and vice versa.

The book consists of two parts. In the first part, we present the theory of the binomial transform for sequences with a sufficient prerequisite of classical numbers and polynomials. The first part provides theorems and tools which help to compute binomial transforms of different sequences and also to generate new binomial identities from old. These theoretical tools (formulas and theorems) can also be used for summation of series and various numerical computations. In the second part, we have compiled a list of binomial transform formulas for easy reference.

In the Appendix, we present the definition of the Stirling sequence transform and a short table of transformation formulas.

Contents

Part 1

Theory of the Binomial Transform

Chapter 1

Introduction

We assume basic knowledge about the binomial coefficients

$$\binom{p}{k} = \frac{p(p-1)...(p-k+1)}{k!}$$

and throughout we use the agreement that $\binom{p}{k} = 0$ if $k < 0$. Here p does not need to be an integer.

Now we define the binomial transform. Given a sequence $\{a_k\}$, $k = 0,1,2...$, its binomial transform is the new sequence $\{b_n\}$, $n = 0,1,2...$ generated by the formula

$$b_n = \sum_{k=0}^{n} \binom{n}{k} a_k \qquad (1.1)$$

with inversion

$$a_n = \sum_{k=0}^{n} \binom{n}{k} (-1)^{n-k} b_k \qquad (1.2)$$

which can be written also as

$$(-1)^n a_n = \sum_{k=0}^{n} \binom{n}{k} (-1)^k b_k.$$

3

The symmetric version of the binomial transform is used very often. In this case the sequence $\{b_k\}$ is defined by

$$b_n = \sum_{k=0}^{n} \binom{n}{k} (-1)^k a_k \tag{1.3}$$

and the inversion formula is

$$a_n = \sum_{k=0}^{n} \binom{n}{k} (-1)^k b_k . \tag{1.4}$$

Here the factor $(-1)^k$ can be replaced by $(-1)^{k-1}$.

A simple computation shows that (1.3) implies the formula

$$\frac{b_{n+1} - a_0}{n+1} = \sum_{k=0}^{n} \binom{n}{k} (-1)^{k+1} \frac{a_{k+1}}{k+1} . \tag{1.5}$$

If the sequence $\{a_k\}$ is indexed from $k = 1, 2, ...$ we can assume that $a_0 = 0$ in order to use the same formulas. In this case we have also $b_0 = 0$.

Here is a proof of the inversion formula (1.2).

$$\sum_{k=0}^{n} \binom{n}{k} (-1)^{n-k} b_k = \sum_{k=0}^{n} \binom{n}{k} (-1)^{n-k} \left\{ \sum_{j=0}^{k} \binom{k}{j} a_k \right\}$$

$$= (-1)^n \sum_{j=0}^{n} a_j \left\{ \sum_{k=j}^{n} \binom{n}{k} \binom{k}{j} (-1)^k \right\} = (-1)^n \sum_{j=0}^{n} a_j (-1)^j \delta_{nj} = a_n .$$

Here the last equality follows from the convolution identity

$$\sum_{k=j}^{n} \binom{n}{k} \binom{k}{j} (-1)^k = (-1)^j \delta_{nj} \tag{1.6}$$

where δ_{nj} is the Kronecker symbol. The inversion formula (1.4) follows from (1.2) as we can write

$$(-1)^n a_n = \sum_{k=0}^n \binom{n}{k}(-1)^{n-k} b_k.$$

The binomial transform naturally appears in the theory of finite differences. Let $\{a_k\}_{k=0}^\infty$ be a sequence and consider the forward difference operator Δ defined by $\Delta a_k = a_{k+1} - a_k$. It is easy to compute that

$$\Delta^n a_0 = \sum_{k=0}^n \binom{n}{k}(-1)^k a_{n-k}$$

and because of the well-known property $\binom{n}{k} = \binom{n}{n-k}$ this can be written also as

$$\Delta^n a_0 = \sum_{k=0}^n \binom{n}{k}(-1)^{n-k} a_k \qquad (1.7)$$

or, equivalently, as

$$(-1)^n \Delta^n a_0 = \sum_{k=0}^n \binom{n}{k}(-1)^k a_k. \qquad (1.8)$$

Another elementary property involves the transform of the shifted sequence. If (1.1) is true then

$$\Delta b_n = b_{n+1} - b_n = \sum_{k=0}^n \binom{n}{k} a_{k+1} \qquad (1.9)$$

for $n \geq 1$ (see [64]). Here is the simple proof

$$b_{n+1} - b_n = \sum_{k=0}^{n+1} \binom{n+1}{k} a_k - \sum_{k=0}^{n} \binom{n}{k} a_k = \sum_{k=0}^{n+1} a_k \left\{ \binom{n+1}{k} - \binom{n}{k} \right\}$$

$$= \sum_{k=1}^{n+1} a_k \binom{n}{k-1} = \sum_{j=0}^{n} a_{j+1} \binom{n}{j}$$

where for the last sum we have set $j = k - 1$. Notice that the summation in (1.1) and (1.9) starts from zero. For correct applications of this property see Example 3.2 in Chapter 3.

Iterating (1.9) we find

$$\sum_{k=0}^{n} \binom{n}{k} a_{k+p} = \Delta^p b_n = \sum_{k=0}^{p} \binom{p}{k} (-1)^{p-k} b_{n+k} \qquad (1.10)$$

for every integer $p \geq 0$. This formula was extended further by Chen [19] to

$$\sum_{k=0}^{n} \binom{n}{k} \binom{p+k}{q} a_{k+p-q} = \sum_{k=0}^{p} \binom{p}{k} (-1)^{p-k} \binom{n+k}{q} b_{n+k-q} .$$

It follows from (1.10) that if we have a recurrence relation

$$a_{n+2} + A a_{n+1} + B a_n = 0$$

(A, B — constants), then it transforms into the difference equation

$$\Delta^2 b_n + A \Delta b_n + B b_n = 0 .$$

Important properties of the binomial transform will be presented in Chapters 4, 5, and 6.

If we iterate (1.1) we find

$$\sum_{m=0}^{n} \binom{n}{m} \left\{ \sum_{k=0}^{m} \binom{m}{k} a_k \right\} = \sum_{k=0}^{n} \binom{n}{k} 2^{n-k} a_k . \qquad (1.11)$$

Formula (1.11) follows from the binomial coefficient property (Knuth, [44, 1.2.6. H])

$$\binom{n}{m}\binom{m}{k} = \binom{n}{k}\binom{n-k}{m-k}. \tag{1.12}$$

The iterated symmetric transform (1.3) is the identity transform, i.e. it leads back to $\{a_k\}$.

There are many identities which can be represented as binomial transforms. For example, the interesting identity

$$\sum_{k=0}^{n} \frac{(-4)^k k!}{(2k+1)!(n-k)!} = \frac{1}{n!(2n+1)} \tag{1.13}$$

proved recently in [27] can be written as a binomial transform formula

$$\sum_{k=0}^{n} \binom{n}{k}(-1)^k \binom{2k}{k}^{-1} \frac{2^{2k}}{2k+1} = \frac{1}{2n+1}$$

(listed as entry (10.39) in Part 2).

Many of the combinatorial identities in [28], [31], [37], [39], and [59] are binomial transforms.

At the same time numerous binomial expressions are not binomial transforms according to our definition. For example, the binomial sums

$$\sum_{k=0}^{n} \binom{n}{k}^2 a_k, \ \sum_{k=0}^{n} \binom{n}{k}\binom{n+k}{m} a_k, \ \sum_{k=0}^{n} \binom{n}{k} a_{2n+k}, \ \sum_{k=0}^{n} \binom{k}{n} a_k$$

are not binomial transforms because they do not fit (1.1) or (1.3).

In this book we present a number of properties of the binomial transform and several theorems which help to compute binomial transforms and also help to generate new formulas and identities from old ones. We have also compiled a table of binomial transform identities which could be used in combinatorics and analysis.

Chapter 2 contains a list of special numbers and polynomials that appear most often in the book. Throughout the text some special numbers and polynomials are introduced at the place for convenience. In Chapter 3 we show in detail the connection of the binomial transform to Euler's series transformation. We present a general series transformation formula of Euler type (Theorem 3.3). The general theorem leads to a number of corollaries, examples and applications. In Chapter 3 we also present the generalized binomial transform. At the end in that chapter we discuss sequences invariant with respect to the binomial transformation.

Chapter 4 is dedicated to the important formula of Melzak

$$\sum_{k=0}^{n} \binom{n}{k} (-1)^k \frac{f(x-k)}{y+k} = \frac{n! \, f(x+y)}{y(y+1)...(y+n)}$$

($f(x)$ is any polynomial with $\deg f \le n$ and $y \ne 0, -1, -2, ..., -n$). The formula was recently extended by the author to arbitrary polynomials. Melzak's formula is a powerful instrument for creating new binomial identities and is also very useful in analysis.

Chapters 5, 6, and 7 contain the main toolbox of the binomial transform theory. Most of these results were obtained recently by the author. In Chapter 5 we show how the image sequence $\{b_n\}$ changes when we multiply of divide the original sequence $\{a_k\}$ by the discrete variable k. In Chapter 6 we study the binomial transform of the product of two sequences. Chapter 7 deals with finite differences

$$\sum_{k=0}^{n} \binom{n}{k} (-1)^k f(x+k) = (-1)^n \Delta^n f(x)$$

and shows their connection to certain series transformation formulas.

Part 2 of the book presents our table of binomial transform formulas. The table is organized as a list of entries with occasional notes which explain the origin of the formula and sometimes its connection to other formulas. The six chapters in the second part contain formulas of different types. Often the inverses are also included. Some formulas are repeated in different chapters for convenience. First they are used to illustrate a certain property or certain technique and then they appear also in the table for completeness. The table contains also some formulas which are not proper binomial transforms, but are considered important.

In Chapter 12 we have included a special section on integration. The purpose is to show how certain trigonometric integrals can be evaluated by using appropriate binomial transform identities.

At the end in the Appendix the reader will find a short table of Stirling transform pairs. The Stirling transform is similar to the binomial transform and there are two Euler type series transformations related to the Stirling transform; namely, (A4) and (A5).

Identities with Stirling numbers appear often in analysis and combinatorics and we believe this short table will be helpful to many.

Chapter 2

Prerequisite: Special Numbers and Polynomials

The binomial transform can be recognized in many formulas involving classical numbers and polynomials. For instance, we have the representation

$$S(\alpha, n) = \frac{1}{n!} \sum_{k=0}^{n} \binom{n}{k} (-1)^{n-k} k^{\alpha} \qquad (2.1)$$

or

$$(-1)^n n! S(\alpha, n) = \sum_{k=0}^{n} \binom{n}{k} (-1)^k k^{\alpha}$$

where $S(\alpha, n), \mathrm{Re}\, \alpha > 0$ are the Stirling functions, the generalized Stirling numbers of the second kind (see [3] for details). When $\alpha = m$ is a positive integer, $S(m, n)$ are the usual Stirling numbers of the second kind [2], [22], [37]. Another notation for these numbers is $S(m, n) = \begin{Bmatrix} m \\ n \end{Bmatrix}$, which is used mostly in combinatorics. The Stirling numbers of the second kind can be defined by the generating function

$$\frac{x^n}{(1-x)(1-2x)...(1-nx)} = \sum_{m=n}^{\infty} S(m, n) x^m \qquad (2.2)$$

11

or by the exponential generating function

$$\frac{1}{n!}(e^x - 1)^n = \sum_{m=n}^{\infty} S(m,n)\frac{x^m}{m!} .$$ (2.3)

The Stirling numbers of the first kind $s(n,k)$ are defined by the generating function

$$x(x-1)...(x-n+1) = n!\binom{x}{n} = \sum_{k=0}^{n} s(n,k)x^k .$$ (2.4)

Their exponential generating function is

$$\frac{\left[\ln(1+x)\right]^k}{k!} = \sum_{n=k}^{\infty} \frac{s(n,k)}{n!} x^n .$$ (2.5)

For the Stirling numbers of both kinds, as well as for other special numbers, good reference are the books of Comtet [22] and Graham *et al.* [37].

Many binomial transform formulas involve the harmonic numbers and the generalized harmonic numbers; for $n = 0, 1, ...$,

$$H_n = \sum_{k=1}^{n} \frac{1}{k}, \quad H_n^{(s)} = \sum_{k=1}^{n} \frac{1}{k^s}, \quad H_0 = H_0^{(s)} = 0,$$ (2.6)

(here s is any complex number). These numbers can be expressed in terms of the digamma (psi) function

$$\psi(x) = \frac{d}{dx}\log\Gamma(x)$$ (2.7)

namely,

$$H_n = \psi(n+1) + \gamma$$ (2.8)

where $\gamma = -\psi(1)$ is Euler's constant. It is good to mention that

$$H_n = \log n + \gamma + O\left(\frac{1}{n}\right). \tag{2.9}$$

When $m \geq 2$ is an integer

$$H_n^{(m)} = \zeta(m) + \frac{(-1)^{m-1}}{(m-1)!} \psi^{(m-1)}(n+1). \tag{2.10}$$

The generating functions of the harmonic numbers and the generalized harmonic numbers are, correspondingly,

$$\frac{-\ln(1-t)}{1-t} = \sum_{n=1}^{\infty} H_n t^n \tag{2.11}$$

$$\frac{1}{1-t} \mathrm{Li}_s(t) = \sum_{n=1}^{\infty} H_n^{(s)} t^n \tag{2.12}$$

where $\mathrm{Li}_s(t)$ is the polylogarithm function

$$\mathrm{Li}_s(t) = \sum_{k=1}^{\infty} \frac{t^k}{k^s}. \tag{2.13}$$

We shall use also the skew-harmonic numbers

$$H_n^- = \sum_{k=1}^{n} \frac{(-1)^{k-1}}{k}, \quad H_0^- = 0 \tag{2.14}$$

with the generating function

$$\frac{\ln(1+t)}{1-t} = \sum_{n=1}^{\infty} H_n^- t^n. \tag{2.15}$$

The Bernoulli polynomials $B_n(x), n = 0, 1, \ldots,$ are very important polynomials in analysis. They are defined by the generating function

$$\frac{te^{xt}}{e^t - 1} = \sum_{n=0}^{\infty} B_n(x) \frac{t^n}{n!} \tag{2.16}$$

and $B_n = B_n(0)$ are the Bernoulli numbers. Similarly, the Euler polynomials $E_n(x), n = 0, 1, \ldots,$ are defined by the generating function

$$\frac{2e^{xt}}{e^t + 1} = \sum_{n=0}^{\infty} E_n(x) \frac{t^n}{n!} \tag{2.17}$$

with $E_n(0) = 0$ when $n = 2, 4, 6, \ldots$ and $E_n(1) = (-1)^n E_n(0)$. The relation

$$E_n(0) = \frac{2}{n+1}(1 - 2^{n+1})B_{n+1}$$

is also helpful.

The Euler numbers are defined by $E_n = 2^n E_n\left(\frac{1}{2}\right)$ or by the generating function

$$\frac{1}{\cosh t} = \sum_{n=0}^{\infty} E_n \frac{t^n}{n!}. \tag{2.18}$$

The Genocchi polynomials are defined by the generating function

$$\frac{2te^{xt}}{e^t + 1} = \sum_{n=0}^{\infty} G_n(x) \frac{t^n}{n!} \tag{2.19}$$

and $G_n = G_n(0)$ are the Genocchi numbers. It is easy to see that $G_n = 2(1 - 2^n)B_n$. Note also that $G_n(x) = nE_{n-1}(x)$.

The Euler-Bernoulli functions $\beta_n(x, \lambda)$ are defined by the generating function

$$\frac{te^{xt}}{\lambda e^t - 1} = \sum_{n=0}^{\infty} \beta_n(x, \lambda) \frac{t^n}{n!}.$$ (2.20)

When $\lambda = 1$, $\beta_n(x, 1) = B_n(x)$ are the Bernoulli polynomials. When $\lambda \neq 1$ and $x = 0$, the functions $\beta_n(\lambda) = \beta_n(0, \lambda)$ are rational function — see [9].

The exponential polynomials $\varphi_n(x), n = 0, 1, \ldots$, appear in several parts of analysis (series summation, asymptotic expansions — see [5], [6], [25]. They are defined by the generating function

$$e^{x(e^t - 1)} = \sum_{n=0}^{\infty} \varphi_n(x) \frac{t^n}{n!}$$

or by the characteristic property

$$\left(x \frac{d}{dx} \right)^n e^x = \varphi_n(x) e^x$$ (2.21)

($n = 0, 1, \ldots$). The coefficients of these polynomials are the Stirling numbers of the second kind

$$\varphi_n(x) = \sum_{k=0}^{n} S(n, k) x^k.$$ (2.22)

The values at $x = 1$

$$\varphi_n(1) = \sum_{k=0}^{n} S(n, k)$$ (2.23)

are the well-known Bell numbers.

The geometric polynomials $\omega_{n,r}(x), n = 0, 1, \ldots,$ $r \geq 0$, where $\omega_{n,0}(x) = \delta_{n,0}$, are defined by the generating function

$$\frac{1}{[1-x(e^t-1)]^r} = \sum_{n=0}^{\infty} \omega_{n,r}(x)\frac{t^n}{n!} \qquad (2.24)$$

or, by the property

$$\sum_{k=0}^{\infty} \binom{k+r-1}{k} k^n x^k = \frac{1}{(1-x)^r} \, \omega_{n,r}\left(\frac{x}{1-x}\right) \qquad (2.25)$$

(see [6], [16]). They have the representation

$$\omega_{n,r}(x) = \frac{1}{\Gamma(r)} \sum_{k=0}^{n} S(n,k)\Gamma(k+r)x^k \qquad (2.26)$$

$$= \sum_{k=0}^{n} S(n,k)\binom{k+r-1}{k} k! x^k.$$

When $r=1$ we write $\omega_{n,1}(x) = \omega_n(x)$ so that

$$\omega_n(x) = \sum_{k=0}^{n} S(n,k)\, k! \, x^k. \qquad (2.27)$$

The numbers

$$\omega_n(1) = \sum_{k=0}^{n} S(n,k)k!$$

are known in combinatorics as the preferential arrangement numbers.
  When $r=1$ we have from (2.25)

$$\sum_{k=0}^{\infty} k^n x^k = \frac{1}{1-x}\, \omega_n\left(\frac{x}{1-x}\right). \qquad (2.28)$$

  The Eulerian polynomials $A_n(x)$ (Comtet, [22], see also [6]) are defined by the equation

$$\sum_{k=0}^{\infty} k^n x^k = \frac{1}{(1-x)^{n+1}} A_n(x) \qquad (2.29)$$

or, by the generating function

$$\frac{1-x}{1-x\exp(t(1-x))} = \sum_{n=0}^{\infty} A_n(x) \frac{t^n}{n!}. \qquad (2.30)$$

The Eulerian polynomials are related to the geometric polynomials by the equation (see [6])

$$\omega_n\left(\frac{x}{1-x}\right) = \frac{A_n(x)}{(1-x)^n}. \qquad (2.31)$$

The Cauchy numbers of first type c_n and the Cauchy numbers of the second type d_n were defined by Comtet [22], pp. 293–294, (see also Jordan [39, pp. 267–279]) as $c_0 = d_0 = 1$ and for $n = 1, 2, ...,$

$$c_n = \int_0^1 x(x-1)...(x-n+1)\,dx \qquad (2.32)$$

$$d_n = \int_0^1 x(x+1)...(x+n-1)\,dx. \qquad (2.33)$$

They have exponential generating functions correspondingly,

$$\frac{t}{\log(1+t)} = \sum_{n=0}^{\infty} c_n \frac{t^n}{n!} \qquad (2.34)$$

$$\frac{-t}{(1-t)\log(1-t)} = \sum_{n=0}^{\infty} d_n \frac{t^n}{n!}. \qquad (2.35)$$

The numbers $(-1)^n d_n$ appeared in the works of Nörlund [54] and are known also as Nörlund numbers. From (2.4) and (2.32), (2.33) we have correspondingly (see [22])

$$c_n = \sum_{k=0}^{n} \frac{s(n,k)}{k+1} \qquad (2.36)$$

$$d_n = \sum_{k=0}^{n} \frac{(-1)^{n-k} s(n,k)}{k+1} \qquad (2.37)$$

where $s(n,k)$ are the Stirling numbers of the first kind (2.4).

The Fibonacci numbers F_n and the Lucas numbers L_n, where $n = 0,1,2,...$, are defined by the generating functions

$$f(t) = \frac{t}{1-t-t^2} = \sum_{k=0}^{\infty} F_k t^k \qquad (2.38)$$

and correspondingly by

$$l(t) = \frac{2-t}{1-t-t^2} = \sum_{k=0}^{\infty} L_k t^k \, . \qquad (2.39)$$

These numbers also have convenient exponential generating functions

$$\frac{e^{\varphi t} - e^{\psi t}}{\varphi - \psi} = \sum_{k=0}^{\infty} \frac{t^k}{k!} F_k \qquad (2.40)$$

$$e^{\varphi t} + e^{\psi t} = \sum_{k=0}^{\infty} \frac{t^k}{k!} L_k \qquad (2.41)$$

where $\varphi = \frac{1}{2}\left(1+\sqrt{5}\right)$, $\psi = \frac{1}{2}\left(1-\sqrt{5}\right) = -1/\varphi$.

The classical Laguerre polynomials $L_n(x), n = 0,1,...$ are defined by the Rodrigues formula

$$L_n(x) = \frac{e^x}{n!}\left(\frac{d}{dx}\right)^n (x^n e^{-x}) \tag{2.42}$$

or by the generating function [58]

$$\frac{1}{1-t}\exp\left(\frac{xt}{1-t}\right) = \sum_{n=0}^{\infty} L_n(x)t^n. \tag{2.43}$$

These polynomials appear as a binomial transform

$$L_n(x) = \sum_{k=0}^{n}\binom{n}{k}\frac{(-x)^k}{k!}. \tag{2.44}$$

(This follows immediately from Euler's series transformation formula — see equation (3.2) below).

The Hermite polynomials $H_n(x)$, $n = 0,1,...$ can be defined by the Rodrigues formula

$$H_n(x) = (-1)^n e^{x^2}\left(\frac{d}{dx}\right)^n e^{-x^2} \tag{2.45}$$

or by the generating function [58], [69]

$$e^{2xt-t^2} = \sum_{n=0}^{\infty} \frac{H_n(x)}{n!}t^n. \tag{2.46}$$

Chapter 3

Euler's Transformation of Series

3.1 Basic Formulas

The binomial transform is closely related to Euler's series transformation. A simple version of this transformation is given in the following proposition.

Proposition 3.1 *Suppose we have a function analytic in a neighborhood of the origin*

$$f(t) = \sum_{k=0}^{\infty} a_k t^k \tag{3.1}$$

or just a formal power series. Euler's series transformation formula says that

$$\frac{1}{1-t} f\left(\frac{t}{1-t}\right) = \sum_{n=0}^{\infty} t^n \left\{ \sum_{k=0}^{n} \binom{n}{k} a_k \right\} \tag{3.2}$$

for $|t|$ small enough.

This follows from the more general Theorem 3.3 below. See also [4] and [30].

Euler's formula can be used, among other things, to evaluate the binomial transform

$$b_n = \sum_{k=0}^{n} \binom{n}{k} a_k \quad (n = 0, 1, 2...) \tag{3.3}$$

by computing the Taylor coefficients of the function on the left hand side in (3.2) independently and comparing coefficients. Examples will be given later.

If we have a convergent series

$$s = f(1) = a_0 + a_1 + ...$$

then with $t = \dfrac{1}{2}$ in (3.2) we find

$$s = \sum_{n=0}^{\infty} \frac{1}{2^{n+1}} \left\{ \sum_{k=0}^{n} \binom{n}{k} a_k \right\}$$

which is Euler's classical formula. Sometimes the new series converges faster, sometimes not — see the examples on pp. 468–470 in [43] and also [62].

With the substitution

$$\frac{t}{1-t} = z, \; t = \frac{z}{z+1}$$

formula (3.2) takes the form

$$f(z) = \frac{1}{z+1} \sum_{n=0}^{\infty} \left(\frac{z}{z+1} \right)^n \left\{ \sum_{k=0}^{n} \binom{n}{k} a_k \right\}. \tag{3.4}$$

The proof of (3.4) and hence of (3.2) follows from the general theorem in the next section. Another proof comes from Proposition 3.14 in Section 3.4.

It is good to mention an interesting formula resulting from (3.4). With x, z appropriate parameters we apply (3.4) to the function

$$f\left(\frac{x}{z}t\right) = \sum_{n=0}^{\infty} \left(\frac{a_n x^n}{z^n}\right) t^n$$

and the results is the representation

$$f(x) = \sum_{n=0}^{\infty} \frac{z^n}{(1+z)^{n+1}} \left\{ \sum_{k=0}^{n} \binom{n}{k} \frac{x^k}{z^k} a_k \right\}. \tag{3.5}$$

Notice that in this representation the variable (parameter) z appears only on the right hand side.

3.2   A General Theorem and Several Transformation Formulas

We shall prove here a general theorem (Theorem 3.3 below) using complex function theory.

Lemma 3.2   *(Hadamard's series multiplication theorem.) Given two power series, say, (3.1) and also a second one*

$$g(t) = \sum_{k=0}^{\infty} c_k t^k \tag{3.6}$$

we have the representation

$$\sum_{n=0}^{\infty} a_n c_n z^n = \frac{1}{2\pi i} \oint_L g\left(\frac{z}{\lambda}\right) f(\lambda) \frac{d\lambda}{\lambda} \tag{3.7}$$

where L is an appropriate closed curve around the origin.
    For details see section 4.6 in [70].
    Now we present our theorem.

Theorem 3.3   *The following representation holds*

$$\sum_{n=0}^{\infty} c_n h(z)^n \left\{ \sum_{k=0}^{n} \binom{n}{k} a_k \right\} = \frac{1}{2\pi i} \oint_L g\left(h(z)\left(1+\frac{1}{\lambda}\right) \right) f(\lambda) \frac{d\lambda}{\lambda} \quad (3.8)$$

where $h(z)$ is an appropriate function for which the above expression is defined and the integral is a Cauchy type integral on a closed curve around the origin, as in (3.7).

Proof. We shall apply Cauchy's integral formula for the coefficients of the function $f(t)$ from (3.1). According to this formula we have for $k = 0,1,...,$

$$a_k = \frac{1}{2\pi i} \oint_L \frac{f(\lambda)d\lambda}{\lambda^{k+1}} \ .$$

From here

$$\sum_{k=0}^{n} \binom{n}{k} a_k = \frac{1}{2\pi i} \oint_L \left(1+\frac{1}{\lambda}\right)^n f(\lambda) \frac{d\lambda}{\lambda} \ . \quad (3.9)$$

Multiplying both sides in this equation by $c_n h(z)^n$ and summing for n we obtain (3.8). Exchanging summation and integration on the right hand side is easily justified. □

Choosing $g(t)$ and $h(z)$ appropriately and combining this result with Hadamard's theorem we shall generate various series trans-formation formulas. Here is the first example.

Corollary 3.4 *Let α be a complex number. Then the following representation holds*

$$\sum_{n=0}^{\infty} \binom{\alpha}{n}(-1)^n a_n z^n \quad (3.10)$$

$$= (z+1)^{\alpha} \sum_{n=0}^{\infty} \left(\frac{z}{z+1}\right)^n \binom{\alpha}{n} (-1)^n \left\{ \sum_{k=0}^{n} \binom{n}{k} a_k \right\}.$$

Proof. In the above theorem we choose

$$h(z) = \frac{z}{z+1}, \quad g(t) = (1-t)^{\alpha} = \sum_{n=0}^{\infty} \binom{\alpha}{n} (-1)^n t^n, \quad c_n = \binom{\alpha}{n} (-1)^n.$$

A simple computation shows that

$$g\left(h(z)\left(1 + \frac{1}{\lambda} \right) \right) = (z+1)^{-\alpha} \left(1 - \frac{z}{\lambda} \right)^{\alpha}$$

and (3.8) takes the form

$$\sum_{n=0}^{\infty} c_n \left(\frac{z}{z+1}\right)^n \left\{ \sum_{k=0}^{n} \binom{n}{k} a_k \right\} = \frac{(1+z)^{-\alpha}}{2\pi i} \oint_L \left(1 - \frac{z}{\lambda} \right)^{\alpha} f(\lambda) \frac{d\lambda}{\lambda}.$$

This representation yields (3.10) in view of Hadamard's theorem. When $\alpha = -1$ we have

$$\binom{-1}{n} = (-1)^n \tag{3.11}$$

and (3.10) becomes (3.4). □

Corollary 3.5 *Let the sequence $\{a_n\}$ be defined by (3.1). Then the following (exponential) version of Euler's series transformation formula holds*

$$\sum_{n=0}^{\infty} \frac{a_n}{n!} z^n = e^{-z} \sum_{n=0}^{\infty} \frac{z^n}{n!} \left\{ \sum_{k=0}^{n} \binom{n}{k} a_k \right\}. \tag{3.12}$$

Proof. In Theorem 3.3 we take $h(z) = z$, $g(t) = e^t$. Then (3.8) becomes

$$\sum_{n=0}^{\infty} \frac{z^n}{n!} \left\{ \sum_{k=0}^{n} \binom{n}{k} a_k \right\} = \frac{e^z}{2\pi i} \oint_L e^{\frac{z}{\lambda}} f(\lambda) \frac{d\lambda}{\lambda}$$

and (3.12) follows from (3.7). $\qquad\qquad\qquad\qquad\qquad\qquad\qquad\square$

This exponential transformation formula can be found in [30] with a direct combinatorial proof.

We can replace in (3.12) a_n by a_n / λ and z by λz (where λ is a parameter) to give the equation the more flexible form

$$e^{\lambda z} \sum_{n=0}^{\infty} \frac{a_n}{n!} z^n = \sum_{n=0}^{\infty} \frac{z^n}{n!} \left\{ \sum_{k=0}^{n} \binom{n}{k} \lambda^{n-k} a_k \right\}. \qquad (3.13)$$

In the next two applications we use the natural logarithmic function. In all expansions we assume that $|z|$ is small enough to secure convergence.

Corollary 3.6 *With $f(t)$ as in (3.1) the following representation holds*

$$a_0 \log(1+z) + \sum_{n=1}^{\infty} \frac{z^n}{n} a_n = \sum_{n=1}^{\infty} \left(\frac{z}{z+1} \right)^n \frac{1}{n} \left\{ \sum_{k=0}^{n} \binom{n}{k} a_k \right\}. \qquad (3.14)$$

Proof. We apply Theorem 3.3 with the choice

$$h(z) = \frac{z}{z+1}, \quad g(t) = -\log(1-t) = \sum_{n=1}^{\infty} \frac{t^n}{n}, \quad c_n = \frac{1}{n}.$$

Then

$$g\left(h(z) \left(1 + \frac{1}{\lambda} \right) \right) = \log(1+z) - \log\left(1 - \frac{z}{\lambda} \right)$$

and the right hand side in (3.8) becomes

$$\frac{\log(1+z)}{2\pi i}\oint_L f(\lambda)\frac{d\lambda}{\lambda} - \frac{1}{2\pi i}\oint_L \log(1-\frac{z}{\lambda})f(\lambda)\frac{d\lambda}{\lambda}.$$

Now (3.14) follows from here. The first term is exactly $a_0\log(1+z)$ and for the second term we use Hadamard's theorem. $\square$

In the next corollary we present an interesting identity involving harmonic numbers.

Corollary 3.7 *For every p with $\operatorname{Re} p > -1$ we have*

$$\sum_{n=0}^{\infty}(H_{p+n}-H_p)\binom{p+n}{n}a_n z^n + \log(1+z)\sum_{n=0}^{\infty}\binom{p+n}{n}a_n z^n \quad (3.15)$$

$$= \frac{1}{(1+z)^{p+1}}\sum_{n=0}^{\infty}\left(\frac{z}{z+1}\right)^n (H_{p+n}-H_p)\binom{p+n}{n}\left\{\sum_{k=0}^{n}\binom{n}{k}a_k\right\}$$

with $\{a_n\}$ an arbitrary sequence of coefficients as in (3.1).

Proof. We use the series expansion (see equation (7.43) in [37], 351–352)

$$g(t) = \frac{-\log(1-t)}{(1-t)^{p+1}} = \sum_{n=0}^{\infty}(H_{p+n}-H_p)\binom{p+n}{n}t^n$$

where $\operatorname{Re} p > -1$. As before we take

$$h(z) = \frac{z}{z+1}$$

and a simple computation gives

$$g\left(h(z)\left(1+\frac{1}{\lambda}\right)\right) = (1+z)^{p+1}\left[\frac{\log(1+z)}{\left(1-\frac{z}{\lambda}\right)^{p+1}} - \frac{\log\left(1-\frac{z}{\lambda}\right)}{\left(1-\frac{z}{\lambda}\right)^{p+1}}\right].$$

Therefore, the right hand side in (3.8) becomes

$$(1+z)^{p+1}\frac{\log(1+z)}{2\pi i}\oint_L \frac{1}{(1-\frac{z}{\lambda})^{p+1}}\frac{f(\lambda)d\lambda}{\lambda}$$

$$-(1+z)^{p+1}\frac{1}{2\pi i}\oint_L \frac{\log(1-\frac{z}{\lambda})^{p+1}}{(1-\frac{z}{\lambda})^{p+1}}\frac{f(\lambda)d\lambda}{\lambda}.$$

At this point we use the well-known binomial expansion (see, for example, [31])

$$\frac{1}{(1-z)^{p+1}} = \sum_{n=0}^{\infty}\binom{p+n}{n}z^n$$

so that according to Hadamard's theorem the first term above becomes

$$(1+z)^{p+1}\log(1+z)\sum_{n=0}^{\infty}\binom{p+n}{n}a_n z^n$$

while the second term is

$$(1+z)^{p+1}\sum_{n=0}^{\infty}(H_{p+n}-H_p)\binom{p+n}{n}a_n z^n$$

and the desired identity follows. In this case we use in (3.8) the coefficients

$$c_n = (H_{p+n} - H_p)\binom{p+n}{n}. \qquad \square$$

The result in Corollary 3.7 was used in [8] to evaluate in closed form the series

$$\sum_{n=0}^{\infty}(H_{p+n} - H_p)\binom{p+n}{n}n^m z^n$$

for any $m \geq 0$ and any $p \geq 0$.

When $p = 0$ then $c_n = H_n$ and we have the special case below.

Corollary 3.8 *With $f(t)$ as in (3.1), the following series transformation formula holds*

$$\sum_{n=0}^{\infty}H_n a_n z^n + \log(1+z)f(z) \qquad (3.16)$$

$$= \frac{1}{1+z}\sum_{n=0}^{\infty}\left(\frac{z}{z+1}\right)^n H_n \left\{\sum_{k=0}^{n}\binom{n}{k}a_k\right\}.$$

For completeness we present here one more series transformation formula involving two power series.

Proposition 3.9 *Given two analytic functions $f(t)$ and $g(t)$ as in (3.1) and (3.6) the following representation is true*

$$\sum_{n=0}^{\infty}a_n c_n t^n = \sum_{n=0}^{\infty}\frac{g^{(n)}(-t)}{n!}t^n\left\{\sum_{k=0}^{n}\binom{n}{k}a_k\right\}. \qquad (3.17)$$

This transformation formula can be found in a modified form in the book [59], Chapter 6, problem 19, p. 245.

Proof. Multiplying both sides in equation (3.9) by

$$\frac{g^{(n)}(-t)}{n!} t^n$$

and summing for n we obtain

$$\sum_{n=0}^{\infty} \frac{g^{(n)}(-t)}{n!} t^n \left\{ \sum_{k=0}^{n} \binom{n}{k} a_k \right\} = \frac{1}{2\pi i} \oint_L \sum_{n=0}^{\infty} \frac{g^{(n)}(-t)}{n!} \left(\frac{t}{\lambda} + t \right)^n f(\lambda) \frac{d\lambda}{\lambda}$$

$$= \frac{1}{2\pi i} \oint_L g\left(\frac{t}{\lambda} \right) f(\lambda) \frac{d\lambda}{\lambda}$$

(recognizing the Taylor expansion of $g\left(\dfrac{t}{\lambda} \right)$ centered at "$-t$" inside the first integral). The result follows now from Hadamard's theorem. □

An application of this formula to series of Hermite polynomials is given in Example 3.6 below.

Euler's transformation works also for asymptotic series. Namely, we have this result:

Corollary 3.10 *Suppose the function*

$$F(\lambda) = \sum_{n=0}^{\infty} \frac{a_n}{\lambda^{n+1}}$$

is analytic in a neighborhood of infinity (or is a formal power series). Then

$$F(\lambda-1) = \sum_{n=0}^{\infty} \frac{1}{\lambda^{n+1}} \left\{ \sum_{k=0}^{n} \binom{n}{k} a_n \right\}.$$

The proof follows easily from (3.2) after the substitution $\lambda = 1/t$ and is left to the reader.

3.3   Examples and Applications

Example 3.1   In this example we show how the transformation formula (3.13) can be used to prove some classical properties of the Bernoulli polynomials $B_n(x)$. In (3.13) we set $a_n = B_n(x)$ and from (2.16)

$$\sum_{n=0}^{\infty} \frac{z^n}{n!} \left\{ \sum_{k=0}^{n} \binom{n}{k} B_k(x)\lambda^{n-k} \right\} = e^{\lambda z} \sum_{n=0}^{\infty} \frac{B_n(x)}{n!} z^n = \frac{ze^{(x+\lambda)z}}{e^z - 1}$$

so that

$$\sum_{n=0}^{\infty} \frac{z^n}{n!} \left\{ \sum_{k=0}^{n} \binom{n}{k} B_k(x)\lambda^{n-k} \right\} = \frac{ze^{(x+\lambda)z}}{e^z - 1} = \sum_{n=0}^{\infty} \frac{z^n}{n!} B_n(x+\lambda) .$$

By comparing coefficients we come to the important identity

$$\sum_{k=0}^{n} \binom{n}{k} B_k(x)\lambda^{n-k} = B_n(x+\lambda) \qquad (3.18)$$

where x, λ are any two numbers. It is known that $B_n(1) = (-1)^n B_n(0) = (-1)^n B_n$, where B_n are the Bernoulli numbers. With $x = 0$ in (3.18) we find the well-known representation

$$\sum_{k=0}^{n} \binom{n}{k} B_k \lambda^{n-k} = B_n(\lambda) \qquad (3.19)$$

and when $\lambda = 1$

$$\sum_{k=0}^{n} \binom{n}{k} B_k = (-1)^n B_n .$$

Note that (3.18) can be written as binomial transform

$$\sum_{k=0}^{n} \binom{n}{k} B_k(x)\lambda^{-k} = \lambda^{-n} B_n(x+\lambda).$$

The above method can be used to compute the binomial transforms of sequences of special numbers or polynomials which have an exponential generating function. For Euler's polynomials $E_n(x)$ with generating function (2.17) we compute in the same way

$$\sum_{k=0}^{n} \binom{n}{k} E_k(x)\lambda^{n-k} = E_n(x+\lambda) \qquad (3.20)$$

by using (3.13). This is true also for the Genocchi polynomials $G_n(x)$ and the Apostol-Bernoulli functions $\beta_n(x,\mu)$ defined in (2.19) and (2.20).

For the exponential polynomials $\varphi_n(x)$ things are somewhat different. Their binomial transform was computed in [5] and will be listed later in Chapter 11.

In the next example and also in several other places we shall use the lemma:

Lemma 3.11   *Given a power series $f(t) = a_0 + a_1 t + a_2 t^2 + ...$, we have*

$$\frac{1}{1-t} f(t) = \sum_{n=0}^{\infty} \left\{ \sum_{k=0}^{n} a_k \right\} t^n. \qquad (3.21)$$

For the proof we expand $\dfrac{1}{1-t} = 1 + t + t^2 + ...$ (assuming for the moment that $|t| < 1$) and use Cauchy's rule for multiplication of two power series. Another way to prove (3.21) is to compute directly the product

$$(1-t)\sum_{n=0}^{\infty} \left\{ \sum_{k=0}^{n} a_k \right\} t^n.$$

Example 3.2 We use now formula (3.2) to find the binomial transform of the sequence $a_n = \dfrac{(-1)^{n-1}}{n}$, $n = 1, 2, \ldots$. The generating function of this sequence is

$$f(t) = \log(1+t) = \sum_{n=1}^{\infty} \frac{(-1)^{n-1}}{n} t^n$$

and then

$$\frac{1}{1-t} f\left(\frac{t}{1-t}\right) = \frac{1}{1-t} \log\left(1 + \frac{t}{1-t}\right) = \frac{1}{1-t} \log\left(\frac{1}{1-t}\right)$$

$$= \frac{-\log(1-t)}{1-t} = \frac{1}{1-t} \sum_{n=1}^{\infty} \frac{t^n}{n} .$$

According to Lemma 3.11 this equals

$$\sum_{n=1}^{\infty} \left(1 + \frac{1}{2} + \ldots + \frac{1}{n}\right) t^n = \sum_{n=1}^{\infty} H_n t^n$$

where H_n are the harmonic numbers (2.6). That is, (3.2) yields

$$\sum_{k=1}^{n} \binom{n}{k} \frac{(-1)^{k-1}}{k} = H_n .\tag{3.22}$$

This well-known formula appears, for instance, on p. 53 of Schwatt's book [61], also on p. 6 in [31] and on p. 5 in [59]. Assuming that the summation starts from zero with $a_0 = 0$ we write

$$\sum_{k=0}^{n} \binom{n}{k} \frac{(-1)^{k-1}}{k} = H_n$$

and then applying (1.9) we find

$$H_{n+1} - H_n = \frac{1}{n+1} = \sum_{j=0}^{n} \binom{n}{j} \frac{(-1)^j}{j+1}. \qquad (3.23)$$

This is an example of a sequence $\dfrac{1}{j+1}$ invariant for the symmetric binomial transform (1.3).

By inversion in (3.22) we have also

$$\sum_{k=1}^{n} \binom{n}{k}(-1)^{k-1} H_n = \frac{1}{n}. \qquad (3.24)$$

Example 3.3   Integrating the representation

$$\frac{-\log(1-t)}{1-t} = \sum_{n=1}^{\infty} H_n t^n$$

and then dividing by t both sides we find the generating function

$$\frac{\log^2(1-t)}{2t} = \sum_{n=1}^{\infty} \frac{H_n}{n+1} t^n. \qquad (3.25a)$$

Replacing here t by $-t$ we obtain also

$$\frac{\log^2(1+t)}{2t} = \sum_{n=1}^{\infty} \frac{(-1)^{n-1} H_n}{n+1} t^n. \qquad (3.25b)$$

We shall use now Euler's series transformation (3.2) to compute the binomial transform of the numbers $a_n = \dfrac{(-1)^{n-1} H_n}{n+1}$. Applying (3.2) to the function in (3.25b) we find

$$\frac{1}{1-t}\left(\frac{1-t}{2t}\right)\log^2\left(\frac{1}{1-t}\right) = \frac{\log^2(1-t)}{2t} = \sum_{n=1}^{\infty} t^n \left\{\sum_{k=0}^{n}\binom{n}{k}\frac{(-1)^{k-1}H_k}{k+1}\right\}$$

and comparing this to the representation (3.25a) we conclude that

$$\sum_{k=0}^{n}\binom{n}{k}\frac{(-1)^{k-1}H_k}{k+1} = \frac{H_n}{n+1} \qquad (3.26)$$

which gives an example of a sequence invariant according to the symmetric binomial transform with factor $(-1)^{k-1}$.

Remark 3.12 *Formula (3.2) can be put in a more flexible equivalent form*

$$\frac{1}{1-\lambda t} f\left(\frac{\mu t}{1-\lambda t}\right) = \sum_{n=0}^{\infty} t^n \left\{\sum_{k=0}^{n}\binom{n}{k}\lambda^{n-k}\mu^k a_k\right\} \qquad (3.27)$$

where λ, μ are appropriate parameters.

To show the equivalence of (3.27) and (3.2) we first write (3.1) in the form

$$f\left(\frac{\mu t}{\lambda}\right) = \sum_{k=0}^{\infty} a_k \left(\frac{\mu}{\lambda}\right)^k t^k$$

and then apply (3.2) to this function as a function of t to get

$$\frac{1}{1-t} f\left(\frac{\mu t}{\lambda(1-t)}\right) = \sum_{n=0}^{\infty} t^n \left\{\sum_{k=0}^{n}\binom{n}{k}\left(\frac{\mu}{\lambda}\right)^k a_k\right\}.$$

Replacing t by λt here we come to (3.27). $\qquad\qquad\square$

Lemma 3.13 *For sufficiently small $|t|$ the following representation holds*

$$\frac{-\log(1-\alpha t)}{1-\beta t} = \sum_{n=1}^{\infty} \left(\alpha\beta^{n-1} + \frac{1}{2}\alpha^2\beta^{n-2} + ... + \frac{1}{n}\alpha^n \right) t^n \quad (3.28)$$

where α and β are appropriate parameters.

Proof. The representation

$$\frac{-\log(1-\alpha t)}{1-t} = \sum_{n=1}^{\infty} \left(\alpha + \frac{1}{2}\alpha^2 + ... + \frac{1}{n}\alpha^n \right) t^n$$

follows from the expansion of the logarithm and Lemma 3.11. We replace here t by βt and then replace $\alpha\beta$ by α to get (3.28). □

Example 3.4   (see [4]) We show that for all λ, μ,

$$\sum_{k=1}^{n} \binom{n}{k} H_k \lambda^{n-k} \mu^k = (\lambda+\mu)^n H_n - \sum_{k=1}^{n} \frac{\lambda^k (\lambda+\mu)^{n-k}}{k}. \quad (3.29)$$

For the proof we apply the transformation formula (3.27) to the function

$$\frac{-\log(1-t)}{1-t} = \sum_{n=1}^{\infty} H_n t^n$$

to find

$$\frac{-1}{1-\lambda t} \frac{\log\left(1-\dfrac{\mu t}{1-\lambda t}\right)}{1-\dfrac{\mu t}{1-\lambda t}} = \frac{-\log(1-(\lambda+\mu)t)}{1-(\lambda+\mu)t} + \frac{\log(1-\lambda t)}{1-(\lambda+\mu)t} \quad (3.30)$$

which equals, in view of (2.11) and the above lemma (with $\alpha = \lambda$ and $\beta = \lambda + \mu$),

$$\sum_{n=1}^{\infty} (\lambda+\mu)^n H_n t^n - \sum_{n=1}^{\infty} \left\{ \sum_{k=1}^{n} \frac{\lambda^k (\lambda+\mu)^{n-k}}{k} \right\} t^n.$$

At the same time (3.27) shows that (3.30) also equals

$$\sum_{n=0}^{\infty} t^n \left\{ \sum_{k=1}^{n} \binom{n}{k} H_k \lambda^{n-k} \mu^k \right\}$$

and comparing coefficients in these power series we finish the proof.
With $\lambda = \mu = 1$ in (3.29), we find a companion to formula (3.24)

$$\sum_{k=1}^{n} \binom{n}{k} H_k = 2^n \left(H_n - \sum_{k=1}^{n} \frac{1}{k 2^k} \right). \tag{3.31}$$

With $\lambda = 1$ in (3.29), we have

$$\sum_{k=1}^{n} \binom{n}{k} H_k \mu^k = (1+\mu)^n H_n - \sum_{k=1}^{n} \frac{(1+\mu)^{n-k}}{k} \tag{3.32}$$

(see also p. 77 in [44]). We apply to this equation the differential
operator $\left(\mu \dfrac{d}{d\mu} \right)^m$ to get

$$\sum_{k=1}^{n} \binom{n}{k} H_k \mu^k k^m = a(m,n,\mu) H_n - \sum_{k=1}^{n} \frac{a(m,n-k,\mu)}{k} \tag{3.33}$$

where

$$a(m,n,\mu) = \left(\mu \frac{d}{d\mu} \right)^m (1+\mu)^n = \sum_{k=0}^{n} \binom{n}{k} k^m \mu^k \tag{3.34}$$

$$= \sum_{k=0}^{n} \binom{n}{k} k! S(m,k) \mu^k (1+\mu)^{n-k}.$$

The second equality here follows from the obvious fact that

$$\left(\mu\frac{d}{d\mu}\right)^m \mu^k = k^m \mu^k .$$

The third equality comes from the representation

$$\sum_{k=0}^{n}\binom{n}{k} k^m \mu^k = \sum_{k=0}^{n}\binom{n}{k} k! S(m,k) \mu^k (1+\mu)^{n-k} \qquad (3.35)$$

which is known and can be found in [31]. It will be given an independent proof later in Example 5.2 in Chapter 5.

Example 3.5   We show here an interesting application of Corollary 3.4 to series with central binomial coefficients. The central binomial coefficients are defined by

$$\binom{2n}{n} = \frac{(2n)!}{(n!)^2} .$$

First we note that by simple computation

$$\binom{-1/2}{n}(-1)^n = \frac{1}{4^n}\binom{2n}{n}$$

and then we set $\alpha = \dfrac{-1}{2}$ in Corollary 3.4 so that formula (3.9) turns into

$$\sum_{n=0}^{\infty}\binom{2n}{n} a_n \frac{z^n}{4^n} = \frac{1}{\sqrt{z+1}}\sum_{n=0}^{\infty}\left(\frac{z}{z+1}\right)^n \frac{1}{4^n}\binom{2n}{n}\left\{\sum_{k=0}^{n}\binom{n}{k} a_k\right\}$$

which after the substitution $z = 4x$ becomes

$$\sum_{n=0}^{\infty}\binom{2n}{n} a_n x^n = \frac{1}{\sqrt{4x+1}}\sum_{n=0}^{\infty}\left(\frac{x}{4x+1}\right)^n \binom{2n}{n}\left\{\sum_{k=0}^{n}\binom{n}{k} a_k\right\} . \qquad (3.36)$$

This transformation formula is interesting by itself!

Setting here

$$a_n = (-1)^{n-1} H_n$$

and using equation (3.23) we find

$$\sum_{n=0}^{\infty} \binom{2n}{n} (-1)^{n-1} H_n x^n = \frac{1}{\sqrt{4x+1}} \sum_{n=1}^{\infty} \left(\frac{x}{4x+1} \right)^n \binom{2n}{n} \frac{1}{n}. \quad (3.37)$$

Now we reach for the well-known expansion (see [12])

$$\sum_{n=0}^{\infty} \binom{2n}{n} \frac{z^n}{n} = 2\ln\left(\frac{1-\sqrt{1-4z}}{2z} \right) = 2\ln\left(\frac{2}{1+\sqrt{1-4z}} \right) \quad (3.38)$$

.

which we use to evaluate in closed form the right hand side in (3.37). Applying (3.38) with

$$z = \frac{x}{4x+1}$$

we find after simple computations

$$\sum_{n=0}^{\infty} \binom{2n}{n} (-1)^{n-1} H_n x^n = \frac{2}{\sqrt{4x+1}} \ln\left(\frac{2\sqrt{4x+1}}{1+\sqrt{4x+1}} \right). \quad (3.39)$$

That is, we have computed the generating function for the numbers

$$\binom{2n}{n} (-1)^{n-1} H_n .$$

This result and several others related to it can be found in [12]. That paper contains also the generating function for the numbers $C_n H_n$ where

$$C_n = \frac{1}{n+1} \binom{2n}{n} \quad (3.40)$$

are the Catalan numbers.

Example 3.6 In this example we present one application of formula (3.17) from Proposition 3.9 to series with Hermite polynomials. Changing t to $-t$ and a_k to $(-1)^k a_k$ in that formula we write it here in the form

$$\sum_{n=0}^{\infty} a_n c_n t^n = \sum_{n=0}^{\infty} \frac{(-1)^n g^{(n)}(t)}{n!} t^n \left\{ \sum_{k=0}^{n} \binom{n}{k} (-1)^k a_k \right\}. \qquad (3.41)$$

Recall now that the Hermite polynomials $H_n(x)$ satisfy the Rodrigues equation

$$H_n(x) = (-1)^n e^{x^2} \left(\frac{d}{dx} \right)^n e^{-x^2}$$

and their generating function is

$$e^{2xt-t^2} = \sum_{n=0}^{\infty} \frac{H_n(x)}{n!} t^n.$$

Corollary 3.14 *The following series transformation formula holds*

$$\sum_{n=0}^{\infty} a_n H_n(x) \frac{t^n}{n!} = e^{2xt-t^2} \sum_{n=0}^{\infty} (-1)^n H_n(x-t) \frac{t^n}{n!} \left\{ \sum_{k=0}^{n} \binom{n}{k} (-1)^k a_k \right\} \quad (3.42)$$

for any sequence of coefficients $\{a_n\}$.

Proof. We use equation (3.41) with $g(t) = e^{2xt-t^2}$ and $c_n = H_n(x)/n!$. From the Rodrigues formula we find

$$\left(\frac{d}{dt} \right)^n e^{2xt-t^2} = \left(\frac{d}{dt} \right)^n e^{x^2} e^{-(x-t)^2}$$

$$= e^{x^2} (-1)^n \left(\frac{d}{dx} \right)^n e^{-(x-t)^2} = e^{x^2} (-1)^n \left(\frac{d}{d(x-t)} \right)^n e^{-(x-t)^2}$$

$$= e^{x^2} (-1)^n (-1)^n e^{-(x-t)^2} H_n(x-t) = e^{2xt-t^2} H_n(x-t) .$$

That is,

$$\left(\frac{d}{dt} \right)^n g(t) = e^{2xt-t^2} H_n(x-t)$$

and (3.42) follows now from (3.41).   □

Next we consider the binomial transform identity

$$(-1)^n n! S(m,n) = \sum_{k=0}^{n} \binom{n}{k} (-1)^k k^m$$

where $S(m,n)$ are the Stirling numbers of the second kind. This is the inversion of (2.1), a well-known analytic representation of the numbers $S(m,n)$. Here m,n are any two nonnegative integers. We substitute $a_k = k^m$ in (3.42) to get the closed form evaluation

$$\sum_{k=0}^{\infty} k^m H_k(x) \frac{t^k}{k!} = e^{2xt-t^2} \sum_{n=0}^{m} S(m,n) H_n(x-t) t^n . \qquad (3.43)$$

The series on the right hand side truncates because $S(m,n) = 0$ for $n > m$.

For more results in this direction see [13].

Example 3.7 The Chebyshev polynomials of the first kind $T_n(x)$ and the Chebyshev polynomials of the second kind $U_n(x)$ have exponential generating functions correspondingly

$$\sum_{n=0}^{\infty} T_n(x)\frac{t^n}{n!} = e^{xt}\cos\left(t\sqrt{1-x^2}\right)$$

$$\sum_{n=1}^{\infty} U_{n-1}(x)\frac{t^n}{n!} = e^{xt}\frac{\sin\left(t\sqrt{1-x^2}\right)}{\sqrt{1-x^2}}$$

(see p. 170 in [69]). Setting $x = \cos\theta$ we can rewrite these series in the form

$$e^{-t\cos\theta}\sum_{n=0}^{\infty} T_n(\cos\theta)\frac{t^n}{n!} = \cos(t\sin\theta)$$

$$e^{-t\cos\theta}\sum_{n=1}^{\infty} U_{n-1}(\cos\theta)\frac{t^n}{n!} = \frac{\sin(t\sin\theta)}{\sin\theta}.$$

Using now Euler's series transformation (3.13) we have

$$\sum_{n=0}^{\infty}\frac{t^n}{n!}\left\{\sum_{k=0}^{n}\binom{n}{k}(-\cos\theta)^{n-k}T_k(\cos\theta)\right\} = \cos(t\sin\theta)$$

$$\sum_{n=0}^{\infty}\frac{t^n}{n!}\left\{\sum_{k=1}^{n}\binom{n}{k}(-\cos\theta)^{n-k}U_{k-1}(\cos\theta)\right\} = \frac{\sin(t\sin\theta)}{\sin\theta}$$

or

$$\sum_{n=0}^{\infty}\frac{t^n}{n!}\left\{(-\cos\theta)^n\sum_{k=0}^{n}\binom{n}{k}(-1)^k\frac{T_k(\cos\theta)}{(\cos\theta)^k}\right\} = \cos(t\sin\theta)$$

$$\sum_{n=0}^{\infty}\frac{t^n}{n!}\left\{(-\cos\theta)^n\sum_{k=1}^{n}\binom{n}{k}(-1)^k\frac{U_{k-1}(\cos\theta)}{(\cos\theta)^k}\right\} = \frac{\sin(t\sin\theta)}{\sin\theta}.$$

Comparing coefficients of both sides we obtain the binomial transforms

$$\sum_{k=0}^{n}\binom{n}{k}(-1)^k\frac{T_k(\cos\theta)}{(\cos\theta)^k}=\begin{cases}0 & (n\text{ odd})\\(-1)^{n/2}(\tan\theta)^n & (n\text{ even})\end{cases}\quad(3.44)$$

$$\sum_{k=1}^{n}\binom{n}{k}(-1)^k\frac{U_{k-1}(\cos\theta)}{(\cos\theta)^k}=\begin{cases}0 & (n\text{ even})\\(-1)^{1+n/2}\dfrac{(\tan\theta)^{n-1}}{\cos\theta} & (n\text{ odd})\end{cases}.$$

Example 3.8 The following representation is true

$$\frac{-2}{1-t}\text{Li}_2\left(\frac{-t}{1-t}\right)=\sum_{n=0}^{\infty}(H_n^2+H_n^{(2)})t^n$$

where Li_2 is the dilogarithm (2.13). We have

$$-\text{Li}_2(-t)=\sum_{n=1}^{\infty}\frac{(-1)^{n-1}t^n}{n^2}.$$

According to Euler's transformation formula (3.1) we have

$$\frac{-1}{1-t}\text{Li}_2\left(\frac{-t}{1-t}\right)=\sum_{n=0}^{\infty}t^n\left(\sum_{k=1}^{n}\binom{n}{k}\frac{(-1)^{k-1}}{k^2}\right)$$

and therefore,

$$\sum_{k=1}^{n}\binom{n}{k}\frac{(-1)^{k-1}}{k^2}=\frac{1}{2}(H_n^2+H_n^{(2)}).$$

This equation will be proved later by a different method — see (9.9) in Part 2.

3.4　Another Extension of Euler's Series Transformation Formula and the Generalized Binomial Transform

Euler's transformation formula (3.27) can be extended in the following way:

Proposition 3.15　*For every* α, λ, μ *and* $f(t) = a_0 + a_1 t + a_2 t^2 + \dots$ *we have*

$$\frac{1}{(1 - \lambda t)^{\alpha+1}} f\left(\frac{\mu t}{1 - \lambda t}\right) = \sum_{n=0}^{\infty} t^n \left\{ \sum_{k=0}^{n} \binom{n+\alpha}{n-k} \lambda^{n-k} \mu^k a_k \right\}. \qquad (3.45)$$

When $\alpha = 0$, this is formula (3.27). Here is a short proof. Starting from the left hand side we write

$$\frac{1}{(1 - \lambda t)^{\alpha+1}} \sum_{k=0}^{\infty} a_k \left(\frac{\mu t}{1 - \lambda t}\right)^k = \sum_{k=0}^{\infty} t^k \mu^k a_k (1 - \lambda t)^{-\alpha-k-1}$$

$$= \sum_{k=0}^{\infty} t^k \mu^k a_k \sum_{m=0}^{\infty} \binom{m+k+\alpha}{m} \lambda^m t^m$$

$$= \sum_{n=0}^{\infty} t^n \left\{ \sum_{k=0}^{n} \binom{n+\alpha}{n-k} \lambda^{n-k} \mu^k a_k \right\}. \qquad \square$$

This transformation formula is related to the *generalized binomial transform*:

Definition　Given a sequence $\{a_k\}$ and a complex number α, the generalized binomial transform, or α-binomial transform $\{b_k = b_k(\alpha)\}$ is the sequence defined by

$$b_n = \sum_{k=0}^{n} \binom{n+\alpha}{n-k} a_k \qquad (3.46)$$

with inversion

$$a_n = \sum_{k=0}^{n} \binom{n+\alpha}{n-k}(-1)^{n-k} b_k \ . \tag{3.47}$$

The symmetrical version is

$$b_n = \sum_{k=0}^{n} \binom{n+\alpha}{n-k}(-1)^k a_k \tag{3.48}$$

with inversion

$$a_n = \sum_{k=0}^{n} \binom{n+\alpha}{n-k}(-1)^k b_k \tag{3.49}$$

see [59], p. 46. Note that Riordan uses in his book the equivalent notation

$$\binom{n+\alpha}{n-k} = \binom{n+\alpha}{k+\alpha} .$$

When $\alpha = 0$ this is the usual binomial transform.

One important example of an α-binomial transform is provided by the generalized Laguerre polynomials

$$L_n^{(\alpha)}(x) = \sum_{k=0}^{n} \binom{n+\alpha}{n-k} \frac{(-x)^k}{k!} \tag{3.50}$$

(see [66]). This equation can be put in the form of an ordinary binomial transform

$$L_n^{(\alpha)}(x) = \frac{\Gamma(n+\alpha+1)}{n!} \sum_{k=0}^{n} \binom{n}{k} \frac{(-x)^k}{\Gamma(k+\alpha+1)} \ .$$

Also, the following identity is useful. For every x

$$\sum_{k=0}^{n}\binom{n+\alpha}{n-k}x^k = \sum_{k=0}^{n}\binom{\alpha+k-1}{k}(1+x)^{n-k} \qquad (3.51)$$

with the agreement that when $\alpha = 0$

$$\binom{k-1}{k} = \begin{cases} 1\,(k=0) \\ 0\,(k>0) \end{cases}$$

(see [7], [52]). In particular,

$$\sum_{k=0}^{n}\binom{n+\alpha}{n-k}(-1)^k = \binom{\alpha+n-1}{n}$$

$$\sum_{k=0}^{n}\binom{n+\alpha}{n-k}\binom{\alpha+k-1}{k}(-1)^k = 1.$$

3.5 Invariant Sequences

A sequence $\{a_n\}$ is invariant for the binomial transform (1.1) when

$$\sum_{k=0}^{n}\binom{n}{k}a_k = a_n \qquad (3.52)$$

for every $n = 0,1,2,...$. In this case we have

$$a_0 = a_0,\ a_0 + a_1 = a_1,\ a_0 + 2a_1 + a_2 = a_2,$$

etc. We see that $a_n = 0$ for every $n = 0,1,2,...$, so this is the zero sequence.

Here is another proof of that. Let

$$g(t) = \sum_{n=0}^{\infty}\frac{t^n}{n!}a_n \qquad (3.53)$$

be the exponential generating function for $\{a_n\}$. From (3.13)

$$e^t \sum_{n=0}^{\infty} \frac{a_n}{n!} t^n = \sum_{n=0}^{\infty} \frac{t^n}{n!} \left\{ \sum_{k=0}^{n} \binom{n}{k} a_k \right\}$$

and we conclude that the sequence $\{a_n\}$ satisfies (3.52) if and only if $e^t g(t) = g(t)$ for every t. This is possible only when $g(t) = 0$ for every t.

If $f(t) = \sum_{k=0}^{\infty} a_k t^k$ is the ordinary generating function for $\{a_n\}$, then according to Euler's formula (3.2) equation (3.52) is equivalent to

$$\frac{1}{1-t} f\left(\frac{t}{1-t}\right) = f(t) . \tag{3.54}$$

Therefore, this functional equation has only trivial solution.

We shall describe now two deferent nontrivial variations. A sequence $\{a_n\}$ has the invariance property

$$\sum_{k=0}^{n} \binom{n}{k} (-1)^k a_k = a_n \tag{3.55}$$

if and only if

$$\frac{1}{1-t} f\left(\frac{-t}{1-t}\right) = f(t) \tag{3.56}$$

(for example, the sequence of Lucas numbers L_n has this property — see equation (5.52) in Chapter 5).

Also, a sequence $\{a_n\}$ satisfies the invariance equation

$$\sum_{k=0}^{n} \binom{n}{k} (-1)^{k-1} a_k = a_n \tag{3.57}$$

if and only if

$$\frac{1}{1-t} f\left(\frac{-t}{1-t}\right) = -f(t) \tag{3.58}$$

(see, for example, the sequence in (3.26) and the Fibonacci sequence (5.45) in Chapter 5).

If $\{a_n\}$ is defined by the exponential generating function (3.53), then:

The sequence $\{a_n\}$ satisfies (3.55) if and only if $e^t g(-t) = g(t)$.

The sequence $\{a_n\}$ satisfies (3.57) if and only if $e^t g(-t) = -g(t)$.

Additional results on invariant sequences can be found in [49], [67], [68], and [75].

Chapter 4

Melzak's Formula and Related Formulas

4.1 The Ordinary Formula of Melzak

Let $n \geq 0$ be an integer and let $f(x)$ be a polynomial of degree $\leq n$. Melzak's formula is the following binomial identity

$$\sum_{k=0}^{n} \binom{n}{k} (-1)^k \frac{f(x-k)}{y+k} = \frac{n! f(x+y)}{y(y+1)...(y+n)} \tag{4.1}$$

true for any $y \neq 0, -1, -2, ..., -n$. This formula was used in the solution of problem 4458 in the American Mathematical Monthly [50] and took its name from one of the authors. However, the formula is much older. It appears, for example, on p. 25 in Nielsen's fundamental study [53]. After the publication of [50] Melzak's formula attracted due attention and was studied and discussed by several authors: see Gould [32], [33], [35], and Prodinger [56]. Gould obtained some interesting extensions, including an extension for polynomials of degree $n+1$. Namely, if $f(x) = a_{n+1}x^{n+1} + a_n x^n + ...$, then

$$\sum_{k=0}^{n} \binom{n}{k} (-1)^k \frac{f(x-k)}{y+k} = \frac{n! f(x+y)}{y(y+1)...(y+n)} - n! a_{n+1} . \tag{4.2}$$

Melzak's formula is very useful for computing binomial transforms. For example, when $f = 1$ in (4.1) we find

$$\sum_{k=0}^{n}\binom{n}{k}(-1)^k\frac{1}{y+k}=\frac{n!}{y(y+1)(y+2)...(y+n)}=\frac{1}{y}\binom{n+y}{n}^{-1} \quad (4.3)$$

or,

$$\sum_{k=0}^{n}\binom{n}{k}(-1)^k\frac{y}{y+k}=\binom{n+y}{n}^{-1} \quad (4.4)$$

and by inversion,

$$\sum_{k=0}^{n}\binom{n}{k}(-1)^k\binom{k+y}{k}^{-1}=\frac{y}{y+k}. \quad (4.5)$$

When $y=1$ in (4.3) we have (see also (3.23))

$$\sum_{k=0}^{n}\binom{n}{k}(-1)^k\frac{1}{k+1}=\frac{1}{n+1} \quad (4.6)$$

or, starting the summation from $k=1$

$$\sum_{k=1}^{n}\binom{n}{k}(-1)^{k-1}\frac{1}{k+1}=\frac{n}{n+1}. \quad (4.7)$$

4.2   Extensions and Applications

The restriction $\deg f \le n$ breaks the symmetry in the binomial transform, so it is important to drop this restriction. We present here an extension of (4.1) to polynomials of any degree (in fact, to any formal power series). We give a short and simple proof of this extension and show that the Stirling numbers of the second kind $S(m,n)$ appear naturally in the new formula.

Before proceeding further, we want to rewrite (4.1) in a more convenient equivalent form. First, without loss of generality we may assume that $x = 0$. Next, using the well-known partial fraction decomposition

$$\frac{n!}{y(y+1)...(y+n)} = \sum_{k=0}^{n}\binom{n}{k}(-1)^k\frac{1}{y+k} \tag{4.8}$$

we write (4.1) in the form

$$\sum_{k=0}^{n}\binom{n}{k}(-1)^k\frac{f(-k)}{y+k} = f(y)\sum_{k=0}^{n}\binom{n}{k}(-1)^k\frac{1}{y+k}$$

or, equivalently,

$$\sum_{k=0}^{n}\binom{n}{k}(-1)^k\frac{f(k)}{k-\lambda} = f(\lambda)\sum_{k=0}^{n}\binom{n}{k}(-1)^k\frac{1}{k-\lambda} \tag{4.9}$$

by replacing y by $-\lambda$ and $f(t)$ by $f(-t)$.

In our theorem we shall use the Stirling numbers of the second kind $S(p,n)$ as defined in (2.1).

Theorem 4.1 *Let $f(t) = a_0 + a_1 t + ...$ be a polynomial (or formal power series). Then for any positive integer n and for any complex number $\lambda \neq 0, 1, 2, ...,$ we have*

$$\sum_{k=0}^{n}\binom{n}{k}(-1)^k\frac{f(k)}{k-\lambda} = \tag{4.10}$$

$$f(\lambda)\sum_{k=0}^{n}\binom{n}{k}\frac{(-1)^k}{k-\lambda} + (-1)^n n! \sum_{m=n+1}^{\infty} a_m \left\{ \sum_{j=0}^{m-1}\lambda^j S(m-j-1,n) \right\}.$$

(The second sum on the right hand side is obviously zero when $f(t)$ is a polynomial of degree $\leq n$.)

Replacing $f(t)$ by $f(a+t)$, where a is an arbitrary number, we can write the formula as

$$\sum_{k=0}^{n}\binom{n}{k}(-1)^{k}\frac{f(a+k)}{k-\lambda}=$$

$$f(a+\lambda)\sum_{k=0}^{n}\binom{n}{k}\frac{(-1)^{k}}{k-\lambda}+(-1)^{n}n!\sum_{m=n+1}^{\infty}\frac{f^{(m)}(a)}{m!}\left\{\sum_{j=0}^{m-1}\lambda^{j}S(m-j-1,n)\right\}.$$

Proof. For the Stirling numbers of the second kind, $S(p,n)$ we have

$$(-1)^{n}n!S(p,n)=\sum_{k=0}^{n}\binom{n}{k}(-1)^{k}k^{p} \tag{4.11}$$

where p and n are any two nonnegative integers. It is well-known that $S(n,n)=1$, $S(n+1,n)=\dfrac{n(n+1)}{2}$, and $S(p,n)=0$ when $p<n$ (see [37]).

Let now m be a positive integer. Obviously,

$$\frac{k^{m}}{k-\lambda}=\frac{k^{m}-\lambda^{m}+\lambda^{m}}{k-\lambda}=k^{m-1}+\lambda k^{m-2}+...+\lambda^{m-1}+\frac{\lambda^{m}}{k-\lambda}.$$

Applying the binomial transform we find

$$\sum_{k=0}^{n}\binom{n}{k}(-1)^{k}\frac{k^{m}}{k-\lambda}$$

$$=\sum_{p=0}^{m-1}\lambda^{p}\left\{\sum_{k=0}^{n}\binom{n}{k}(-1)^{k}k^{m-p-1}\right\}+\lambda^{m}\sum_{k=0}^{n}\binom{n}{k}(-1)^{k}\frac{1}{k-\lambda}.$$

In view of (4.11) this equation becomes

$$\sum_{k=0}^{n}\binom{n}{k}(-1)^{k}\frac{k^{m}}{k-\lambda}$$

$$= (-1)^n n! \sum_{p=0}^{m-1} \lambda^p S(m-p-1,n) + \lambda^m \sum_{k=0}^{n} \binom{n}{k} (-1)^k \frac{1}{k-\lambda}.$$

We now multiply both sides by the coefficient a_m and sum for $m = 1, 2, \ldots$. Adding to both sides also the sum

$$\sum_{k=0}^{n} \binom{n}{k} (-1)^k \frac{a_0}{k-\lambda}$$

we obtain the desired formula, namely,

$$\sum_{k=0}^{n} \binom{n}{k} (-1)^k \frac{f(k)}{k-\lambda}$$

$$= f(\lambda) \sum_{k=0}^{n} \binom{n}{k} \frac{(-1)^k}{k-\lambda} + (-1)^n n! \sum_{m=n+1}^{\infty} a_m \left\{ \sum_{p=0}^{m-1} \lambda^p S(m-p-1,n) \right\}.$$

The summation in the second sum on the right hand side starts from $m = n+1$, because when $m < n+1$ we have $m-1 < n$ and $S(m-1-p,n) = 0$. The proof is complete. □

When $f(t)$ is of degree $\leq n$, the sum on the right hand side in (4.10) is zero and we have the original formula of Melzak (4.1). When the degree of $f(t)$ equals $n+1$, then $a_m = 0$ for $m > n+1$ and (4.10) turns into

$$\sum_{k=0}^{n} \binom{n}{k} (-1)^k \frac{f(k)}{k-\lambda} = f(\lambda) \sum_{k=0}^{n} \binom{n}{k} \frac{(-1)^k}{k-\lambda} + (-1)^n n! a_{n+1}$$

in accordance with (4.2) (remembering the replacement of $f(t)$ by $f(-t)$). When the degree of $f(t)$ is $n+2$ we have

$$\sum_{k=0}^{n} \binom{n}{k} (-1)^k \frac{f(k)}{k-\lambda}$$

$$= f(\lambda)\sum_{k=0}^{n}\binom{n}{k}\frac{(-1)^k}{k-\lambda}+(-1)^n n!\left(a_{n+1}+a_{n+2}\left(\frac{n(n+1)}{2}+\lambda\right)\right)$$

etc.   .

Remark 4.2   *Formula (4.10) can be viewed from a different perspective by using the representation*

$$\sum_{k=0}^{n}\binom{n}{k}\frac{(-1)^k}{k-\lambda}=\frac{(-1)^{n+1}n!}{\lambda(\lambda-1)...(\lambda-n)}=(-1)^{n+1}n!\sum_{m=n}^{\infty}\frac{S(m,n)}{\lambda^{m+1}} \quad (4.12)$$

where the infinite series converges for $|\lambda|>n$.
(See (7.47) in [37].)

Corollary 4.3 *For any polynomial or formal power series $f(t)=a_0+a_1t+...$ and every $n=1,2,...,$ we have*

$$\sum_{k=1}^{n}\binom{n}{k}(-1)^{k-1}\frac{f(k)}{k} \quad (4.13)$$

$$= f'(0)+f(0)H_n+(-1)^{n-1}n!\sum_{m=n+1}^{\infty}a_m S(m-1,n)$$

where H_n are the harmonic numbers. In particular, when $\deg f\le n$

$$\sum_{k=1}^{n}\binom{n}{k}(-1)^{k-1}\frac{f(k)}{k}=f'(0)+f(0)H_n. \quad (4.14)$$

Proof. We consider first the case when $f(t)$ has degree $\le n$ and apply formula (4.9). On both sides we separate the term with $k=0$ to write

$$\frac{f(0)}{-\lambda}+\sum_{k=1}^{n}\binom{n}{k}(-1)^k\frac{f(k)}{k-\lambda}=\frac{f(\lambda)}{-\lambda}+f(\lambda)\sum_{k=1}^{n}\binom{n}{k}(-1)^k\frac{1}{k-\lambda}$$

then we bring the first term on the left to the right hand side and multiply both sides by -1 to get

$$\sum_{k=1}^{n}\binom{n}{k}(-1)^{k-1}\frac{f(k)}{k-\lambda} = \frac{f(\lambda)-f(0)}{\lambda} + f(\lambda)\sum_{k=1}^{n}\binom{n}{k}(-1)^{k-1}\frac{1}{k-\lambda}\ .$$

Setting here $\lambda \to 0$ we find

$$\sum_{k=1}^{n}\binom{n}{k}(-1)^{k-1}\frac{f(k)}{k} = f'(0) + f(0)\sum_{k=1}^{n}\binom{n}{k}(-1)^{k-1}\frac{1}{k}$$

which in view of identity (3.22), that is,

$$\sum_{k=1}^{n}\binom{n}{k}(-1)^{k-1}\frac{1}{k} = H_n$$

becomes equation (4.14). The general case follows from (4.10). The first two sums in (4.10) we manipulate as above and then with $\lambda \to 0$ the last sum in (4.10) becomes

$$(-1)^{n-1}n!\sum_{m=n+1}^{\infty} a_m\, S(m-1,n)\ .$$

The proof is completed. $\qquad\qquad\qquad\qquad\qquad\qquad\qquad\square$

Remark 4.4　*When we apply (4.14) to the polynomial $f(t)=t^{p+1}$, we find*

$$\sum_{k=1}^{n}\binom{n}{k}(-1)^{k-1}k^{p} = 0, \qquad\qquad (4.15)$$

for any $p<n$. In the case $p=n$

$$\sum_{k=1}^{n} \binom{n}{k} (-1)^{k-1} k^n = (-1)^{n-1} n! \ . \tag{4.16}$$

This phenomenon was discussed in [2] (see also Remark 7.2 in Chapter 7). For $p \geq n$ we have from (4.13)

$$\sum_{k=1}^{n} \binom{n}{k} (-1)^{k-1} k^p = (-1)^{n-1} n! \, S(p,n) \ .$$

This is exactly the representation (2.1).

Formula (4.14) can be found in Henry Gould's treasure trove of identities [31].

Remark 4.5   *From formula (4.2) we can derive the following identity*

$$\sum_{k=0}^{n} \binom{n}{k} (-1)^k \frac{f(x-k)}{(y+k)(z+k)} = \tag{4.17}$$

$$\frac{n!}{z-y} \left\{ \frac{f(x+y)}{y(y+1)...(y+n)} - \frac{f(x+z)}{z(z+1)...(z+n)} \right\}$$

for any polynomial $f(x)$ with $\deg f \leq n+1$ and variables $y \neq z$.

For the proof all we need is the decomposition

$$\frac{1}{(y+k)(z+k)} = \frac{1}{z-y} \left\{ \frac{1}{y+k} - \frac{1}{z+k} \right\} \ .$$

The degree of the polynomial $f(x)$ can be increased to $n+1$, because the constant $n! a_{n+1}$ in (4.2) cancels out.

(See pp. 83–85 in [31] and also [35].)

Corollary 4.6 *Given* $f(t) = a_0 + a_1 t + ...$, *for every integer* $n > 0$, *and every* $\lambda \neq 0, 1, 2, ...$, *we have*

$$\sum_{k=0}^{n} \binom{n}{k} (-1)^k \frac{f(k)}{(k-\lambda)^2} = \qquad (4.18)$$

$$f'(\lambda) \sum_{k=0}^{n} \binom{n}{k} \frac{(-1)^k}{k-\lambda} + f(\lambda) \sum_{k=0}^{n} \binom{n}{k} \frac{(-1)^k}{(k-\lambda)^2}$$

$$+ (-1)^n n! \sum_{m=n+1}^{\infty} a_m \left\{ \sum_{j=1}^{m-1} j \lambda^{j-1} S(m-j-1, n) \right\}.$$

This follows from Theorem 4.1 after differentiation with respect to λ.

Further differentiation in (4.18) provides a formula for

$$\sum_{k=0}^{n} \binom{n}{k} (-1)^k \frac{f(k)}{(k-\lambda)^3}$$

etc.

Chapter 5

Special Properties. Creating New Identities

5.1   Multiplication and Division by the Discrete Variable

Given the binomial transform (1.1), we want to see what happens when we multiply or divide the original sequence $\{a_k\}$ by the discrete variable k.

The two classical integral transforms, the Laplace transform and the Fourier transform, have the important property that they convert, roughly speaking, multiplication by the variable into differentiation. We shall see here that the discrete binomial transform has a similar property — it converts multiplication by the discrete variable k into the operator $n\nabla$ where ∇ is the backward difference.

Introducing the backward difference operator $\nabla b_n = b_n - b_{n-1}$ we have the following theorem.

Theorem 5.1   *Suppose we have the transform*

$$\sum_{k=0}^{n}\binom{n}{k} a_k = b_n \tag{5.1}$$

then for every positive integer p and every $n \geq p$

$$\sum_{k=0}^{n}\binom{n}{k} k^p a_k = (n\nabla)^p b_n . \tag{5.2}$$

Here we need $n \geq p$ in the above formula because the right hand side contains a term with b_{n-p} which is not defined for $n < p$. We shall discuss this situation later. It is clear that we only need to prove the property for $p = 1$ as the rest follows by iteration. The proof will be given at the end of the chapter.

We have the immediate corollary.

Corollary 5.2 *Let* $g(t)$ *be a polynomial. Then*

$$\sum_{k=0}^{n} \binom{n}{k} g(k) a_k = g(n\nabla) b_n .$$

In particular, for every complex number λ *and every integer* $n \geq 1$,

$$\sum_{k=0}^{n} \binom{n}{k} (k + \lambda) a_k = (n\nabla + \lambda) b_n = (n + \lambda) b_n - n b_{n-1} . \qquad (5.3)$$

By inversion in (5.2) we have:

Corollary 5.3 *For every* $n \geq 1$

$$\sum_{k=1}^{n} \binom{n}{k} (-1)^{n-k} k \nabla a_k = n b_n \qquad (5.4)$$

or

$$\sum_{k=1}^{n} \binom{n}{k} (-1)^{k} k \nabla a_k = (-1)^{n} n b_n .$$

The next corollary describes division by $k + \lambda$.

Corollary 5.4 *Let* $a_0 = 0$ *(so that* $b_0 = 0$ *too). Then for every number* $\lambda \neq -1, -2, \ldots,$ *and every* $n \geq 1$,

$$\sum_{k=1}^{n}\binom{n}{k}\frac{a_k}{k+\lambda} = \sum_{m=1}^{n}\frac{(m+1)(m+2)...n}{(\lambda+m)(\lambda+m+1)...(\lambda+n)}b_m \qquad (5.5)$$

$$= n!\sum_{m=1}^{n}\frac{b_m}{m!(\lambda+m)(\lambda+m+1)...(\lambda+n)}.$$

When $\lambda = 0$ equation (5.5) takes the form

$$\sum_{k=1}^{n}\binom{n}{k}\frac{a_k}{k} = \sum_{m=1}^{n}\frac{b_m}{m}. \qquad (5.6)$$

Property (5.6) was proved by A.N. 't Woord in [77] by a different method. When $a_0 = 0$, the two properties (5.2) and (5.6) are, in fact, equivalent. We shall prove this later, at the end of the chapter.

The case $\lambda = 1$ in (5.5) is also interesting,

$$\sum_{k=1}^{n}\binom{n}{k}\frac{a_k}{k+1} = \frac{1}{n+1}\sum_{m=1}^{n}b_m. \qquad (5.7)$$

When $a_0 \neq 0, b_0 \neq 0$ we can apply (5.5) to the transform

$$\sum_{k=1}^{n}\binom{n}{k}a_k = b_n - b_0$$

and avoid the restriction. Thus we get, for instance, (with summation on both sides from zero)

$$\sum_{k=0}^{n}\binom{n}{k}\frac{a_k}{k+1} = \frac{1}{n+1}\sum_{m=0}^{n}b_m.$$

In the next section we present a number of examples. We want to demonstrate that the above properties are very effective tools for evaluating binomial transforms and for generating various binomial identities. All proofs (for the theorem and the corollaries) will be given at the end of the chapter.

The above properties are true, of course, if we use the symmetric version of the transform with the alternating factor $(-1)^{k-1}$ or $(-1)^k$ in (5.1). That is, if

$$\sum_{k=0}^{n} \binom{n}{k} (-1)^{k-1} a_k = b_n$$

then

$$\sum_{k=0}^{n} \binom{n}{k} (-1)^{k-1} k^p a_k = (n\nabla)^p b_n \qquad (5.8)$$

etc.

5.2   Examples

Here we use the above results in order to obtain several new binomial transform identities and to give short proofs to some known ones. Our examples will involve harmonic, Stirling, and Fibonacci numbers, and also Laguerre polynomials.

Theorem 5.1 can be used in two different ways: computing the binomial transform of $k^p a_k$ by computing $(n\nabla)^p b_n$ or computing $(n\nabla)^p b_n$ if the left hand side in (5.2) can be evaluated by other means.

Example 5.1   We start with something simple. In the alternating version (5.8) consider the sequence $a_k = 1, k = 0,1,...,$ where $b_0 = -1, b_n = 0$ $(n \geq 1)$. We have

$$\sum_{k=0}^{n} \binom{n}{k} (-1)^{k-1} = 0 \ (n \geq 1) . \qquad (5.9)$$

Then for $p = 1$,

$$\sum_{k=0}^{n} \binom{n}{k} (-1)^{k-1} k = \begin{cases} 0 \ (n \neq 1) \\ 1 \ (n=1) \end{cases}. \tag{5.10}$$

In general, for any positive integer p

$$\sum_{k=0}^{n} \binom{n}{k} (-1)^{k-1} k^{p} = (-1)^{n-1} n! S(p,n) \tag{5.11}$$

where $S(p,n)$ are the Stirling numbers of the second kind with the property $S(p,n) = 0$ when $n > p$ and $S(n,n) = 1$.

Example 5.2   Consider the identity

$$\sum_{k=0}^{n} \binom{n}{k} x^{k} = (1+x)^{n}.$$

Here $a_{k} = x^{k}$, where x is any real or complex number and $b_{n} = (1+x)^{n}$. We can write

$$\sum_{k=0}^{n} \binom{n}{k} k^{p} x^{k} \tag{5.12}$$

$$= \left(x \frac{d}{dx} \right)^{p} (1+x)^{n} = \sum_{j=0}^{p} S(p,j) x^{j} \left(\frac{d}{dx} \right)^{j} (1+x)^{n}$$

$$= \sum_{j=0}^{p} S(p,j) x^{j} n(n-1)...(n-j+1)(1+x)^{n-j}$$

$$= \sum_{j=0}^{p} \binom{n}{j} S(p,j) \, j! \, x^{j} (1+x)^{n-j}.$$

Here we are using the following differentiation rule (see, for example, [5] or [59, p. 218]). For any p-times differentiable function $f(x)$

$$\left(x\frac{d}{dx}\right)^p f(x) = \sum_{j=0}^{p} S(p,j)x^j \left(\frac{d}{dx}\right)^j f(x).$$

The theorem then implies the formula: For every for $n \geq p$

$$(n\nabla)^p (1+x)^n = \sum_{j=0}^{p} \binom{n}{j} S(p,j)\, j!\, x^j (1+x)^{n-j}.$$

Example 5.3   Now we shall use property (5.6). Consider the sequence $a_0 = 0$, $a_k = 1\,(n \geq 1)$. Then $b_0 = 0$ and for $n \geq 1$

$$\sum_{k=1}^{n} \binom{n}{k}(-1)^{k-1}1 = 1.$$

From (5.6) we get the well-known identity

$$\sum_{k=1}^{n} \binom{n}{k}(-1)^{k-1}\frac{1}{k} = 1 + \frac{1}{2} + ... + \frac{1}{n} = H_n \qquad (5.13)$$

where H_n are the harmonic numbers. This is another proof of (3.22). Repeating (5.6) we find also

$$\sum_{k=1}^{n} \binom{n}{k}(-1)^{k-1}\frac{1}{k^2} = \sum_{k=1}^{n}\frac{H_k}{k} \qquad (5.14)$$

etc. Notice that by a simple computation

$$\sum_{k=1}^{n}\frac{H_k}{k} = \frac{1}{2}\left(H_n^2 + H_n^{(2)}\right)$$

where $H_n^{(2)} = 1 + \frac{1}{2^2} + ... + \frac{1}{n^2}$.

By inversion in (5.14) we find

$$\sum_{k=1}^{n}\binom{n}{k}(-1)^{k-1}\left(H_k^2 + H_k^{(2)}\right) = \frac{2}{n^2} \tag{5.15}$$

and again by (5.6)

$$\sum_{k=1}^{n}\binom{n}{k}(-1)^{k-1}\frac{1}{k}\left(H_k^2 + H_k^{(2)}\right) = 2\sum_{k=1}^{n}\frac{1}{k^3} = 2H_n^{(3)}. \tag{5.16}$$

Inverting this formula we get

$$\sum_{k=1}^{n}\binom{n}{k}(-1)^{k-1}H_n^{(3)} = \frac{1}{2n}\left(H_n^2 + H_n^{(2)}\right). \tag{5.17}$$

Using now (5.2) we find for all $n > 1$

$$\sum_{k=1}^{n}\binom{n}{k}(-1)^{k-1}k\left(H_n^2 + H_n^{(2)}\right) = n\nabla\frac{2}{n^2} = \frac{2(1-2n)}{n(n-1)^2} \tag{5.18}$$

$$\sum_{k=1}^{n}\binom{n}{k}(-1)^{k-1}k^2\left(H_n^2 + H_n^{(2)}\right) \tag{5.19}$$

$$= n\nabla\left(\frac{2(1-2n)}{n(n-1)^2}\right) = \frac{4}{(n-1)(n-2)} \quad (n>2)$$

$$\sum_{k=1}^{n}\binom{n}{k}(-1)^{k-1}k^3\left(H_n^2 + H_n^{(2)}\right) \tag{5.20}$$

$$= n\nabla\left(\frac{4}{(n-1)(n-2)}\right) = \frac{8n(n-2)}{(n-1)(n-3)} \quad (n>3)$$

etc. At the same time, inverting (5.13) we have

$$\sum_{k=1}^{n}\binom{n}{k}(-1)^{k-1}H_k = \frac{1}{n} \tag{5.21}$$

and (5.6) yields

$$\sum_{k=1}^{n}\binom{n}{k}(-1)^{k-1}\frac{H_k}{k} = H_n^{(2)} . \tag{5.22}$$

Inverting this we have

$$\sum_{k=1}^{n}\binom{n}{k}(-1)^{k-1} H_k^{(2)} = \frac{H_n}{n} \tag{5.23}$$

and (5.2) yields consecutively

$$\sum_{k=1}^{n}\binom{n}{k}(-1)^{k-1} k H_k^{(2)} = n\nabla\left(\frac{H_n}{n}\right) = \frac{1-H_n}{n-1} \tag{5.24}$$

$$\sum_{k=1}^{n}\binom{n}{k}(-1)^{k-1} k^2 H_k^{(2)} = n\nabla\left(\frac{1-H_n}{n-1}\right) = \frac{1-2n+nH_n}{(n-1)(n-2)} \tag{5.25}$$

etc. Subtracting (5.23) from (5.15) we find

$$\sum_{k=1}^{n}\binom{n}{k}(-1)^{k-1} H_k^2 = \frac{2}{n^2} - \frac{H_n}{n} . \tag{5.26}$$

Likewise, subtracting (5.24) from (5.18) gives

$$\sum_{k=1}^{n}\binom{n}{k}(-1)^{k-1} k H_k^2 = \frac{H_n}{n-1} + \frac{2-3n-n^2}{n(n-1)^2} . \tag{5.27}$$

Proceeding further we can evaluate

$$\sum_{k=1}^{n}\binom{n}{k}(-1)^{k-1} k^p H_k^2$$

for higher values of p. Chuanan Wei *et al.* have evaluated these sums up to $p = 3$ in [76] by a different method.

Example 5.4 We continue to work here with the identity (5.14)

$$\sum_{k=1}^{n}\binom{n}{k}(-1)^{k-1}\frac{1}{k^2}=\sum_{k=1}^{n}\frac{H_k}{k}$$

from the previous example. It can be written also this way

$$\sum_{k=1}^{n}\binom{n}{k}(-1)^{k-1}\frac{1}{k^2}=\sum\frac{1}{k_1 k_2}, \quad 1\le k_1\le k_2\le n . \tag{5.28}$$

Applying property (5.6) yields

$$\sum_{k=1}^{n}\binom{n}{k}(-1)^{k-1}\frac{1}{k^3}=\sum\frac{1}{k_1 k_2 k_3} \tag{5.29}$$

where the sum is for $1\le k_1\le k_2\le k_3\le n$. Continuing this process we find for every positive integer m

$$\sum_{k=1}^{n}\binom{n}{k}\frac{(-1)^{k-1}}{k^m}=\sum\frac{1}{k_1 k_2 ... k_m} \tag{5.30}$$

with summation for $1\le k_1\le k_2\le ...\le k_m\le n$.

This formula was obtained by Karl Dilcher in [26] by other means (see also [4] for a different proof).

Example 5.5 Here we shall use property (5.2). Consider the sequence of harmonic numbers $a_k = H_k, n=0,1,...$ with $a_0 = 0, b_0 = 0$. For $n>0$ we have from (5.21)

$$\sum_{k=0}^{n}\binom{n}{k}(-1)^{k-1}H_k=\frac{1}{n}$$

(summation can start from $k = 0$). Therefore, from (5.2) it follows that for any $n > 1$

$$\sum_{k=0}^{n}\binom{n}{k}(-1)^{k-1}kH_k = n\left(\frac{1}{n}-\frac{1}{n-1}\right) = \frac{-1}{n-1} \qquad (5.31)$$

and for $n > 2$,

$$\sum_{k=0}^{n}\binom{n}{k}(-1)^{k-1}k^2 H_k = n\left(\frac{-1}{n-1}+\frac{1}{n-2}\right) = \frac{n}{(n-1)(n-2)} \qquad (5.32)$$

etc. In fact, we have the general formula (3.33). We can write it in the form

$$\sum_{k=1}^{n}\binom{n}{k}H_k k^p x^k = a(p,n,x)H_n - \sum_{k=1}^{n-1}\frac{a(p,k,x)}{n-k} \qquad (5.33)$$

where

$$a(p,n,x) = (x\frac{d}{dx})^p (x+1)^n \qquad (5.34)$$

$$= \sum_{j=0}^{n}\binom{n}{j}S(p,j)j!x^j(1+x)^{n-j}.$$

With $x = -1$ we find $a(p,n,-1) = (-1)^n n! S(p,n)$ and thus

$$\sum_{k=1}^{n}\binom{n}{k}(-1)^{k-1}H_k k^p \qquad (5.35)$$

$$= (-1)^{n-1}n!S(p,n)H_n + \sum_{k=1}^{n-1}\frac{(-1)^k k!S(p,k)}{n-k}.$$

The formula is true for any positive p and n. When $n > p$ the first term on the right hand side is missing, since $S(p,n) = 0$. In view of our theorem this formula explains the action of the operator $(n\nabla)^p$ on the sequence $\{1/n\}_{n=1}^{\infty}$ for $n > p$, that is,

$$(n\nabla)^p \frac{1}{n} = \sum_{k=1}^{n-1} \frac{(-1)^k k! S(p,k)}{n-k}. \qquad (5.36)$$

Example 5.6 Starting from

$$\sum_{k=1}^{n} \binom{n}{k}(-1)^{k-1} x^k = 1 - (1-x)^n \qquad (5.37)$$

we find by inversion

$$\sum_{k=1}^{n} \binom{n}{k}(-1)^{k-1} \left[1 - (1-x)^k\right] = x^n \qquad (5.38)$$

and then according to (5.6)

$$\sum_{k=1}^{n} \binom{n}{k}(-1)^{k-1} \left\{ \frac{1-(1-x)^k}{k} \right\} = \sum_{k=1}^{n} \frac{x^k}{k}. \qquad (5.39)$$

With $x = -1$ we get

$$\sum_{k=1}^{n} \binom{n}{k}(-1)^{k-1} \left\{ \frac{1-2^k}{k} \right\} = \sum_{k=1}^{n} \frac{(-1)^k}{k} = -H_n^- \qquad (5.40)$$

where

$$H_n^- = 1 - \frac{1}{2} + \dots + \frac{(-1)^{n-1}}{n} \quad (n \geq 1), \ H_0^- = 0$$

are the skew-harmonic numbers.

From (5.40) by inversion,

$$\sum_{k=1}^{n}\binom{n}{k}(-1)^{k-1}H_k^- = \frac{2^n-1}{n} \tag{5.41}$$

and according to (5.2)

$$\sum_{k=1}^{n}\binom{n}{k}(-1)^{k-1}kH_k^- = n\nabla\left\{\frac{2^n-1}{n}\right\} \tag{5.42}$$

$$= 2^{n-1}\frac{n-2}{n-1}+\frac{1}{n-1} .$$

Applying (5.6) to (5.41) yields

$$\sum_{k=1}^{n}\binom{n}{k}(-1)^{k-1}\frac{H_k^-}{k} = \sum_{k=1}^{n}\frac{2^k}{k^2} - H_n^{(2)} \tag{5.43}$$

and with (5.7) applied to (5.41),

$$\sum_{k=1}^{n}\binom{n}{k}(-1)^{k-1}\frac{H_k^-}{k+1} = \frac{1}{n+1}\left\{\sum_{k=1}^{n}\frac{2^k}{k} - H_n\right\} . \tag{5.44}$$

Example 5.7 The operator $n\nabla$ works very well on sequences defined by recurrence relations. For example, let $F_0, F_1, F_2,...$ be the sequence of Fibonacci numbers defined by $F_0=0, F_1=1$, and $F_n=F_{n-1}+F_{n-2}$ (the books [45] and [73] are good references). We shall prove some new identities for these numbers. It is known that for every $n\geq 0$,

$$\sum_{k=0}^{n}\binom{n}{k}(-1)^{k-1}F_k = F_n \tag{5.45}$$

and also

$$\sum_{k=0}^{n} \binom{n}{k} F_k = F_{2n}. \tag{5.46}$$

Here is a simple proof. The Fibonacci numbers have the representation (Binet's formula)

$$F_n = \frac{1}{\sqrt{5}} \left(\varphi^n - \psi^n \right) \tag{5.47}$$

where φ is the golden ratio and ψ is its conjugate

$$\varphi = \frac{1+\sqrt{5}}{2}, \; \psi = \frac{1-\sqrt{5}}{2}.$$

Both these numbers are solutions to the quadratic equation $\varphi^2 = \varphi + 1$, so that we have $\varphi + \psi = 1$ and $\varphi\psi = -1$.

Now we write

$$\sum_{k=0}^{n} \binom{n}{k} F_k = \frac{1}{\sqrt{5}} \left(\sum_{k=0}^{n} \binom{n}{k} \varphi^k - \sum_{k=0}^{n} \binom{n}{k} \psi^k \right)$$

$$= \frac{1}{\sqrt{5}} \left((1+\varphi)^n - (1+\psi)^n \right) = \frac{1}{\sqrt{5}} (\varphi^{2n} - \psi^{2n}) = F_{2n}$$

$$\sum_{k=0}^{n} \binom{n}{k} (-1)^{k-1} F_k = \frac{-1}{\sqrt{5}} \left(\sum_{k=0}^{n} \binom{n}{k} (-\varphi)^k - \sum_{k=0}^{n} \binom{n}{k} (-\psi)^k \right)$$

$$= \frac{-1}{\sqrt{5}} \left((1-\varphi)^n - (1-\psi)^n \right) = \frac{-1}{\sqrt{5}} (\psi^n - \varphi^n) = F_n.$$

Applying (5.2) or (5.8) to (5.45) we find

$$\sum_{k=0}^{n} \binom{n}{k} (-1)^{k-1} k F_k = n(F_n - F_{n-1}) = nF_{n-2} \tag{5.48}$$

and further

$$\sum_{k=0}^{n}\binom{n}{k}(-1)^{k-1}k^2 F_k = n^2 F_{n-4} + n F_{n-3} \qquad (5.49)$$

etc. From (5.46) we find correspondingly

$$\sum_{k=0}^{n}\binom{n}{k}k F_k = n(F_{2n} - F_{2n-2}) = n F_{2n-1} \qquad (5.50)$$

$$\sum_{k=0}^{n}\binom{n}{k}k^2 F_k = n^2 F_{2n-2} + n F_{2n-3} \qquad (5.51)$$

etc. Almost the same equations are satisfied by the Lucas numbers L_n as they have the representation $L_n = \varphi^n + \psi^n$. In the same way we compute

$$\sum_{k=0}^{n}\binom{n}{k}(-1)^k L_k = L_n \qquad (5.52)$$

$$\sum_{k=0}^{n}\binom{n}{k}L_k = L_{2n}. \qquad (5.53)$$

The Lucas numbers satisfy the same recurrence $L_n = L_{n-1} + L_{n-2}$ and therefore,

$$\sum_{k=0}^{n}\binom{n}{k}(-1)^k k L_k = n(L_n - L_{n-1}) = n L_{n-2} \qquad (5.54)$$

$$\sum_{k=0}^{n}\binom{n}{k}k L_k = n(L_{2n} - L_{2n-2}) = n L_{2n-1} \qquad (5.55)$$

etc. For the Fibonacci numbers we also have from (5.7)

$$\sum_{k=0}^{n} \binom{n}{k} (-1)^{k-1} \frac{F_k}{k+1} = \frac{1}{n+1} \sum_{k=0}^{n} F_k = \frac{1}{n+1} (F_{n+2} - 1) \quad (5.56)$$

by using the fundamental property $F_1 + F_2 + ... + F_n = F_{n+2} - 1$ and the fact that $F_0 = 0$. For the Lucas numbers we also have $L_0 + L_1 + L_2 + ... + L_n = L_{n+2} - 1$ and therefore,

$$\sum_{k=0}^{n} \binom{n}{k} (-1)^k \frac{L_k}{k+1} = \frac{1}{n+1} (L_{n+2} - 1). \quad (5.57)$$

Example 5.8   We shall use now a recurrence property to evaluate one special binomial transform. Let q be a nonnegative integer. For $n \geq 1$ we define

$$\sigma_n(q) = 1^q + 2^q + ... + n^q \quad (5.58)$$

and also $\sigma_0(q) = 0$. We want to compute the sequence

$$b_n = \sum_{k=0}^{n} \binom{n}{k} (-1)^{k-1} \sigma_k(q) \quad (n = 0,1,...).$$

Obviously, $b_0 = 0$ and $b_1 = 1$. Next we invert the above equation to get

$$\sum_{k=0}^{n} \binom{n}{k} (-1)^{k-1} b_k = \sigma_n(q)$$

and using property (5.2) for $n \geq 1$

$$\sum_{k=0}^{n} \binom{n}{k} (-1)^{k-1} k b_k = n(\sigma_n(q) - \sigma_{n-1}(q)) = n^{q+1} .$$

At the same time, by inversion in (5.11)

$$\sum_{k=0}^{n} \binom{n}{k} k! S(q+1, k) = n^{q+1} . \quad (5.59)$$

Therefore, $(-1)^{k-1}kb_k = k!S(q+1,k)$ and $b_k = (-1)^{k-1}(k-1)!S(q+1,k)$ for $k = 1, 2, \dots$. That is,

$$\sum_{k=0}^{n}\binom{n}{k}(-1)^{k-1}\sigma_k(q) = (-1)^{n-1}(n-1)!S(q+1,n) \qquad (5.60)$$

for $n \ge 1$. It is remarkable that this sequence truncates, $b_n = 0$ for $n > q+1$.

Example 5.9   This example is related to the previous one. For $q \ge 1$ we can write equation (5.59) in the form

$$\sum_{k=1}^{n}\binom{n}{k}k!S(q,k) = n^q \qquad (5.61)$$

with summation from $k = 1$, since $S(q,0) = 0$ for $q \ge 1$. Now (5.7) provides the new formula

$$\sum_{k=1}^{n}\binom{n}{k}\frac{k!S(q,k)}{k+1} = \frac{1}{n+1}\sum_{k=1}^{n}k^q = \frac{\sigma_n(q)}{n+1} . \qquad (5.62)$$

Example 5.10   We shall use again equation (5.21)

$$\sum_{k=1}^{n}\binom{n}{k}(-1)^{k-1}H_k = \frac{1}{n} \ (n \ge 1)$$

and apply property (5.5). Thus for any $\lambda \ne -1, -2, \dots$, we have

$$\sum_{k=1}^{n}\binom{n}{k}(-1)^{k-1}\frac{H_k}{k+\lambda} = \sum_{m=1}^{n}\frac{(m+1)(m+2)\dots n}{(\lambda+m)(\lambda+m+1)\dots(\lambda+n)m} . \qquad (5.63)$$

For $\lambda = 1$ this turns into the symmetric formula

$$\sum_{k=1}^{n} \binom{n}{k} (-1)^{k-1} \frac{H_k}{k+1} = \frac{H_n}{n+1} \tag{5.64}$$

which is equation (3.26) in Example 3.3.

For $\lambda = 2, 3, 4$ in (5.63) we have correspondingly.

$$\sum_{k=1}^{n} \binom{n}{k} (-1)^{k-1} \frac{H_k}{k+2} = \frac{H_n + n}{(n+1)(n+2)} \tag{5.65}$$

$$\sum_{k=1}^{n} \binom{n}{k} (-1)^{k-1} \frac{H_k}{k+3} = \frac{n^2 + 7n + 4H_n}{2(n+1)(n+2)(n+3)} \tag{5.66}$$

$$\sum_{k=1}^{n} \binom{n}{k} (-1)^{k-1} \frac{H_k}{k+4} = \frac{2n^3 + 21n^2 + 85n + 36H_n}{6(n+1)(n+2)(n+3)(n+4)}. \tag{5.67}$$

(For some comments on these results see [10].)

Example 5.11 Consider the sequence $a_k = k$, $k = 1, 2, ...$, where $b_1 = 1$ and $b_n = 0$ for $n > 1$ (see Example 5.1). Suppose $\lambda \neq 0, -1, -2, ...$ We apply property (5.5) to the equation

$$\sum_{k=1}^{n} \binom{n}{k} (-1)^{k-1} k = \begin{cases} 0 \ (n > 1) \\ 1 \ (n = 1) \end{cases} \tag{5.68}$$

to get

$$\sum_{k=1}^{n} \binom{n}{k} (-1)^{k-1} \frac{k}{k+\lambda} = \frac{n!}{(\lambda+1)(\lambda+2)...(\lambda+n)}. \tag{5.69}$$

From here, writing $\dfrac{k}{k+\lambda} = \dfrac{k+\lambda-\lambda}{k+\lambda} = 1 - \dfrac{\lambda}{k+\lambda}$ we find

$$\frac{1}{\lambda} + \sum_{k=1}^{n} \binom{n}{k} (-1)^k \frac{1}{k+\lambda} = \frac{n!}{\lambda(\lambda+1)(\lambda+2)...(\lambda+n)}$$

or, starting the summation from $k = 0$,

$$\sum_{k=0}^{n} \binom{n}{k} (-1)^k \frac{1}{k+\lambda} = \frac{n!}{\lambda(\lambda+1)(\lambda+2)...(\lambda+n)} \ . \tag{5.70}$$

This well-known identity follows also from Melzak's formula (see equation (4.3) in Chapter 4).

Example 5.12 Let $L_n(x) = \dfrac{e^x}{n!} \left(\dfrac{d}{dx}\right)^n (x^n e^{-x})$ be the Laguerre polynomials. According to (2.44) we have

$$\sum_{k=0}^{n} \binom{n}{k} \frac{(-x)^k}{k!} = L_n(x). \tag{5.71}$$

This can be written also in the form (starting the summation from $k = 1$)

$$\sum_{k=1}^{n} \binom{n}{k} \frac{(-x)^k}{k!} = L_n(x) - 1. \tag{5.72}$$

Using property (5.6) we find a new identity

$$\sum_{k=1}^{n} \binom{n}{k} \frac{(-x)^k}{k!k} = \sum_{k=1}^{n} \frac{L_k(x)}{k} - H_n \ . \tag{5.73}$$

The left hand side here can be constructed also by dividing equation (5.72) by the variable x and integrating. Thus

$$\sum_{k=1}^{n} \binom{n}{k} \frac{(-x)^k}{k!k} = \int_0^x \frac{L_k(t)-1}{t} dt \ . \tag{5.74}$$

Comparing the last two identities we conclude that (cf. [8])

$$\sum_{k=1}^{n} \frac{L_k(x)}{k} = \int_0^x \frac{L_k(t)-1}{t} dt + H_n .$$

(5.75)

Next, applying (5.7) to (5.72) we find

$$\sum_{k=1}^{n} \binom{n}{k} \frac{(-x)^k}{k!(k+1)} = \frac{1}{n+1} \sum_{k=1}^{n} (L_k(x)-1)$$

(5.76)

and then by integration in (5.72) we obtain the property

$$\frac{1}{n+1} \sum_{k=1}^{n} (L_k(x)-1) = \frac{1}{x} \int_0^x L_n(t) - 1 \, dt .$$

(5.77)

Example 5.13 Let p be a nonnegative integer. Consider the binomial transform

$$\sum_{k=0}^{n} \binom{n}{k} \binom{p}{k} = \binom{p+n}{p}$$

(5.78)

which is known as the Vandermonde identity [31], [37]. According to (5.2) we have

$$\sum_{k=0}^{n} \binom{n}{k} \binom{p}{k} k = n \left\{ \binom{p+n}{p} - \binom{p+n-1}{p} \right\}$$

(5.79)

$$= n \binom{p+n-1}{p-1} .$$

Applying (5.7) to (5.78) we find also

$$\sum_{k=0}^{n} \binom{n}{k} \binom{p}{k} \frac{1}{k+1} = \frac{1}{n+1} \sum_{j=0}^{n} \binom{p+j}{p}$$

(5.80)

$$= \frac{1}{n+1}\binom{p+n+1}{p+1} .$$

5.3 Proofs

For the proof of Theorem 5.1 we only need to show that if for every $n \geq 0$ we have

$$\sum_{k=0}^{n}\binom{n}{k}a_k = b_n$$

then for $n \geq 1$

$$\sum_{k=0}^{n}\binom{n}{k}ka_k = n(b_n - b_{n-1}) .$$

First we notice that for any $n \geq 1$ and any $0 \leq k \leq n$

$$\binom{n}{k} - \binom{n-1}{k} = \frac{k}{n}\binom{n}{k} .$$

Therefore, we have

$$b_n - b_{n-1} = \sum_{k=0}^{n}\left[\binom{n}{k} - \binom{n-1}{k}\right]a_k = \sum_{k=0}^{n}\left[\frac{k}{n}\binom{n}{k}\right]a_k = \frac{1}{n}\sum_{k=0}^{n}\binom{n}{k}ka_k$$

and the assertion follows for $p = 1$. Repeating this $p-1$ times we obtain (5.2). ☐

Proof of Corollary 5.4 Let $\lambda \neq -1, -2, \ldots$. We define the sequence $\{c_n\}_{n=1}^{\infty}$ by the equation

$$\sum_{k=1}^{n}\binom{n}{k}\frac{a_k}{k+\lambda} = c_n .$$

When $n = 1$ we find directly from here $c_1 = \dfrac{b_1}{1 + \lambda}$. For $n > 1$ we find from (5.3)

$$\sum_{k=1}^{n} \binom{n}{k} a_k = (n + \lambda)c_n - nc_{n-1} = b_n$$

and we can compute c_n from the recurrence relation

$$(n + \lambda)c_n - nc_{n-1} = b_n$$

for $n \geq 2$. Thus for $n = 2$ we have $(\lambda + 2)c_2 - 2c_1 = b_2$ and therefore,

$$c_2 = \frac{2b_1}{(\lambda + 1)(\lambda + 2)} + \frac{b_2}{\lambda + 2}.$$

For $n = 3$ we find in the same manner

$$c_3 = \frac{2 \cdot 3}{(\lambda + 1)(\lambda + 2)(\lambda + 3)}b_1 + \frac{3}{(\lambda + 2)(\lambda + 3)}b_2 + \frac{1}{\lambda + 3}b_3$$

and the result follows by induction. The details are left to the reader. □

Remark 5.5 *When $a_0 = 0$ (and hence $b_0 = 0$) properties (5.2) and (5.6) are equivalent.*

Proof. We just need to see that (5.2) follows from (5.6). Here we shall give the proof. Suppose (5.6) is true. Then for $n = 1, 2, ...,$ we define the sequence c_n by

$$\sum_{k=1}^{n} \binom{n}{k} k a_k = c_n.$$

Clearly, $c_1 = b_1$. For $n > 1$ we find from (5.6)

$$\sum_{k=1}^{n}\binom{n}{k}a_k = \sum_{m=1}^{n}\frac{c_m}{m} = b_n$$

and now we can compute c_n in terms of b_n and b_{n-1} from the recurrence relation

$$\sum_{m=1}^{n}\frac{c_m}{m} = b_n \ . \tag{5.81}$$

For $n = 2$ we have $\dfrac{c_2}{2} + c_1 = b_2$, that is, $c_2 = 2(b_2 - b_1)$. It is easy to see that if (5.81) holds and $c_m = m(b_m - b_{m-1})$ is true for $m \le n$, then it is also true for $m = n + 1$. The proof follows by induction. The simple algebra is left to the reader.   □

Chapter 6

Binomial Transforms of Products

6.1 The Main Theorem

Let $\{a_n\}$ be a sequence and let its binomial transform be the sequence $\{b_n\}$, that is,

$$b_n = \sum_{k=0}^{n} \binom{n}{k} a_k \tag{6.1}$$

with inverse formula

$$a_n = \sum_{k=0}^{n} \binom{n}{k} (-1)^{n-k} b_k \ .$$

Our purpose in this chapter is to develop a technique that helps to generate special new binomial identities from known identities. When the binomial transform (6.1) is given, we want to compute the sequence

$$\sum_{k=0}^{n} \binom{n}{k} a_k c_k \tag{6.2}$$

$(n = 0, 1, ...)$, where the transform

$$d_n = \sum_{k=0}^{n} \binom{n}{k} (-1)^{n-k} c_k, \quad c_n = \sum_{k=0}^{n} \binom{n}{k} d_k \tag{6.3}$$

is also known.

The following theorem is true:

Theorem 6.1   *Let $\{a_n\}$ and $\{c_n\}$ be two sequences and let $\{b_n\}$ and $\{d_n\}$ be defined by (6.1) and (6.3). Then we have the identity*

$$\sum_{k=0}^{n}\binom{n}{k} a_k c_k = \sum_{m=0}^{n}\binom{n}{m} d_m \nabla^m b_n \tag{6.4}$$

where $\nabla b_n = b_n - b_{n-1}$ with $\nabla^0 b_n = b_n$.

For the symmetric version of the binomial transform $\{c_n\} \leftrightarrow \{d_n\}$, namely,

$$d_n = \sum_{k=0}^{n}\binom{n}{k}(-1)^k c_k, \quad c_n = \sum_{k=0}^{n}\binom{n}{k}(-1)^k d_k$$

equation (6.4) takes the form

$$\sum_{k=0}^{n}\binom{n}{k} a_k c_k = \sum_{m=0}^{n}\binom{n}{m}(-1)^m d_m \nabla^m b_n . \tag{6.5}$$

As we shall see, with appropriate choices of the sequences $\{a_n\}$ and $\{c_n\}$ this formula produces interesting new identities involving polynomials and special numbers. In several cases the iterated differences $\nabla^m b_n$ can be computed explicitly.

The proof of the theorem is based on the following lemma.

Lemma 6.2   *Suppose the sequences $\{a_n\}$ and $\{b_n\}$ are given by (6.1) and let $\nabla b_n = b_n - b_{n-1}$, $\nabla^0 b_n = b_n$. Then for every two integers $0 \le m \le n$ we have*

$$\sum_{k=0}^{n}\binom{n}{k} k(k-1)...(k-m+1) a_k = m!\binom{n}{m}\nabla^m b_n \tag{6.6}$$

or, in a shorter form

$$\sum_{k=0}^{n}\binom{n}{k}\binom{k}{m}a_k =\binom{n}{m}\nabla^m b_n .\qquad (6.7)$$

For $m=1$ this is the property

$$\sum_{k=0}^{n}\binom{n}{k}ka_k =n\nabla b_n \qquad (6.8)$$

which was established in the previous chapter.

Note that the summation in (6.6) and (6.7) starts, in fact, from $k=m$, as all previous terms are zeros.

The next lemma shows one possible way to compute the right hand side in (6.6).

Lemma 6.3 *For any two integers $0\le m\le n$ we have*

$$\binom{n}{m}\nabla^m b_n =\sum_{j=0}^{n}\binom{n}{j}\binom{j}{n-m}(-1)^{n-j}b_j .$$

This identity can also be written in the form

$$\nabla^m b_n = m! \sum_{j=n-m}^{n}\frac{(-1)^{n-j}b_j}{(n-j)!(j-n+m)!} . \qquad (6.9)$$

The proofs of the two lemmas and the theorem are given at the end of the chapter in Section 6.5.

Example 6.1 Let $a_k =1$ for all $k\ge 0$. Then

$$b_n =\sum_{k=0}^{n}\binom{n}{k}=2^n$$

and Lemma 6.1 implies that for any $n = 0, 1, 2, \ldots$, and any $0 \le m \le n$

$$\sum_{k=0}^{n} \binom{n}{k} k(k-1)\ldots(k-m+1) = m! \binom{n}{m} \nabla^m 2^n \qquad (6.10)$$

$$= m! \binom{n}{m} 2^{n-m}$$

as by a simple computation we find $\nabla^m 2^n = 2^{n-m}$.

In the next section we apply Theorem 6.1 to the case $c_k = x^k$, and in Section 6.3 we apply the results to special numbers. Among other things we shall work with the sequence $a_k = (-1)^{k-1} H_k$, where H_k are the harmonic numbers. The iterated differences $\nabla^m H_n$ will be computed and this will provide the identity

$$\sum_{k=0}^{n} \binom{n}{k} (-1)^{k-1} H_k c_k = (-1)^{n-1} H_n d_n + \sum_{m=0}^{n-1} \frac{(-1)^m d_m}{n-m}$$

for any sequence $\{c_k\}$ where c_k and d_k are related by (6.3). We shall use this formula when the terms c_k are the Fibonacci numbers, the Bernoulli numbers, and the Laguerre polynomials.

Another similar identity proved in Section 6.3 involves the Fibonacci numbers

$$\sum_{k=0}^{n} \binom{n}{k} (-1)^{k-1} F_k c_k = \sum_{m=0}^{n} \binom{n}{m} d_m F_{n-2m}$$

with c_k and d_k as above.

6.2   Binomial Polynomials

For a given sequence $\{a_n\}$ we consider the polynomials

$$p_n(x) = \sum_{k=0}^{n} \binom{n}{k} a_k x^k$$

which are called here binomial polynomials. When the binomial transform (6.1) is known we want to compute the polynomials $p_n(x)$ explicitly in terms of the numbers $p_n(1) = b_n$. We present here two solutions to this problem. They both follow from Theorem 6.1 with different choices of the sequences $\{a_n\}$ and $\{c_n\}$.

Corollary 6.4 *Assuming the binomial transform (6.1) is given, we have the identity*

$$\sum_{k=0}^{n} \binom{n}{k} a_k x^k = \sum_{j=0}^{n} \binom{n}{j} b_j x^j (1-x)^{n-j} \qquad (6.11)$$

$$= (1-x)^n \sum_{j=0}^{n} \binom{n}{j} b_j \left(\frac{x}{1-x}\right)^j.$$

Proof. In Theorem 6.1 we set

$$a_n = (-1)^n x^n = \sum_{k=0}^{n} \binom{n}{k} (-1)^{n-k} (1-x)^k$$

so that in view of (6.1), $b_n = (1-x)^n$. Simple computation shows that for any number z and any two non-negative integers m and n we have

$$\nabla^m z^n = z^{n-m} (z-1)^m$$

and thus

$$\nabla^m b_n = \nabla^m (1-x)^n = (1-x)^{n-m} (-x)^m.$$

Now from Theorem 6.1

$$\sum_{k=0}^{n}\binom{n}{k}(-1)^{k}x^{k}c_{k} = \sum_{m=0}^{n}\binom{n}{m}d_{m}(-x)^{m}(1-x)^{n-m}$$

$$= \sum_{m=0}^{n}\binom{n}{m}(-1)^{m}d_{m}\,x^{m}(1-x)^{n-m}\,.$$

Here we change notations in order to write this equation in terms of $\{a_{n}\}$ and $\{b_{n}\}$. Setting $a_{k} = (-1)^{k}c_{k}$ we have from (6.3)

$$(-1)^{n}d_{n} = \sum_{k=0}^{n}\binom{n}{k}(-1)^{k}c_{k} = \sum_{k=0}^{n}\binom{n}{k}a_{k} = b_{n}$$

and the above equation becomes

$$\sum_{k=0}^{n}\binom{n}{k}a_{k}x^{k} = \sum_{m=0}^{n}\binom{n}{m}b_{m}x^{m}(1-x)^{n-m}$$

as needed.   □

Second proof independent of Theorem 6.1.   Using   the   inversion
formula we can write

$$p_{n}(x) = \sum_{k=0}^{n}\binom{n}{k}a_{k}x^{k} = \sum_{k=0}^{n}\binom{n}{k}x^{k}\left\{\sum_{j=0}^{k}\binom{k}{j}(-1)^{k-j}b_{j}\right\}$$

$$= \sum_{j=0}^{n}(-1)^{j}b_{j}\left\{\sum_{k=j}^{n}\binom{n}{k}\binom{k}{j}(-1)^{k}x^{k}\right\}$$

and the result follows from the well-known identity from Gould's collection, [31, (3.11.8)]

$$\sum_{k=j}^{n}\binom{n}{k}\binom{k}{j}(-1)^{k}x^{k} = (-1)^{j}\binom{n}{j}x^{j}(1-x)^{n-j}. \qquad (6.12)$$

                                                                             □

Remark 6.5   *Identity (6.12) itself follows from Corollary 6.4 applied to the convolution formula*

$$\sum_{k=j}^{n} \binom{n}{k}\binom{k}{j}(-1)^k = (-1)^n \delta_{nj} . \tag{6.13}$$

Next we consider some examples of representations like (6.11).

Example 6.2   The generalized Stirling numbers $S(\alpha,n)$ of the second kind are defined by the binomial formula (2.1), that is,

$$\sum_{k=0}^{n} \binom{n}{k}(-1)^k k^\alpha = (-1)^n n! S(\alpha,n)$$

where α is any complex number with $\mathrm{Re}\,\alpha > 0$. According to Corollary 6.4 we have

$$\sum_{k=0}^{n} \binom{n}{k}(-1)^k k^\alpha x^k = \sum_{j=0}^{n} \binom{n}{j} j! S(\alpha,j)(-1)^j x^j (1-x)^{n-j}$$

or changing x to $-x$

$$\sum_{k=0}^{n} \binom{n}{k} k^\alpha x^k = \sum_{j=0}^{n} \binom{n}{j} j! S(\alpha,j) x^j (1+x)^{n-j} . \tag{6.14}$$

Example 6.3   Setting $x=2$ in (6.11) we obtain the curious identity

$$\sum_{k=0}^{n} \binom{n}{k} 2^k a_k = \sum_{j=0}^{n} \binom{n}{j}(-1)^{n-j} 2^j b_j .$$

Also, with $x=1/2$ in (6.11) we find

$$\sum_{k=0}^{n} \binom{n}{k} 2^{n-k} a_k = \sum_{j=0}^{n} \binom{n}{j} b_j$$

which is identity (1.11).

Example 6.4   In this example we apply Corollary 6.4 to formula (5.70) from the previous chapter, namely,

$$\sum_{k=0}^{n}\binom{n}{k}(-1)^k\frac{1}{k+\lambda}=\frac{n!}{\lambda(\lambda+1)(\lambda+2)...(\lambda+n)}$$

where $\lambda \neq 0,-1,-2,...$. The result is

$$\sum_{k=0}^{n}\binom{n}{k}\frac{(-1)^k x^k}{k+\lambda} \qquad\qquad (6.15)$$

$$=(1-x)^n\sum_{j=0}^{n}\binom{n}{j}\frac{j!}{\lambda(\lambda+1)(\lambda+2)...(\lambda+j)}\left(\frac{x}{1-x}\right)^j$$

$$=\sum_{j=0}^{n}\binom{n}{j}\frac{j!x^j(1-x)^{n-j}}{\lambda(\lambda+1)(\lambda+2)...(\lambda+j)} \ .$$

With $x=1$ this becomes (5.70), as only the term with $j=n$ remains in the sum. With $x=-1$ in (6.15) we also find

$$\sum_{k=0}^{n}\binom{n}{k}\frac{1}{k+\lambda}=\sum_{j=0}^{n}\binom{n}{j}\frac{(-1)^j 2^{n-j} j!}{\lambda(\lambda+1)(\lambda+2)...(\lambda+j)} \ .$$

Setting $x=1/2$ and $\lambda=m$ (a positive integer) in (6.15) yields

$$\sum_{k=0}^{n}\binom{n}{k}\frac{(-1)^k}{2^k(k+m)}=\frac{1}{2^n}\sum_{j=0}^{n}\binom{n}{j}\frac{j!}{m(m+1)(m+2)...(m+j)}$$

$$=\frac{n!(m-1)!}{2^n(n+m)!}\sum_{k=0}^{n}\binom{n+m}{k}$$

which is the identity obtained by Helmut Prodinger (*Quaest. Math.*, 32 (2009), 265–267).

The representation (6.11) is short and simple, but its right hand side is not a polynomial in standard form. From Theorem 6.1 we obtain also a second corollary:

Corollary 6.6 *Suppose the sequence $\{b_n\}$ is the binomial transform of the sequence $\{a_n\}$. Then*

$$\sum_{k=0}^{n}\binom{n}{k}a_k x^k = \sum_{m=0}^{n}\binom{n}{m}\nabla^m b_n (x-1)^m . \tag{6.16}$$

Proof. Taking $c_k = x^k$ in (6.3) we have the one-line proof

$$d_n = \sum_{k=0}^{n}\binom{n}{k}(-1)^{n-k} x^k = (-1)^n \sum_{k=0}^{n}\binom{n}{k}(-x)^k = (x-1)^n$$

and (6.16) follows from (6.4).   □

Formula (6.16) gives, in fact, the Taylor expansion of the polynomial $p_n(x)$ centered at $x=1$. Many existing examples of binomial polynomials have this form.

Example 6.5   With $x=2$ in (6.16) we find

$$\sum_{k=0}^{n}\binom{n}{k}2^k a_k = \sum_{m=0}^{n}\binom{n}{m}\nabla^m b_n .$$

In view of Example 6.3 we also have

$$\sum_{m=0}^{n}\binom{n}{m}\nabla^m b_n = \sum_{j=0}^{n}\binom{n}{j}(-1)^{n-j} 2^j b_j .$$

Example 6.6   This is one very simple demonstration. When $a_n = (-1)^{n-1}$ we have for all $n \geq 0$

$$\sum_{k=0}^{n}\binom{n}{k}(-1)^{k-1} = \begin{cases} -1, n = 0 \\ 0, n > 0 \end{cases}$$

and from (6.9) we find $\nabla^m b_n = 0$ for $m \neq n$ and $\nabla^n b_n = (-1)^{n-1}$. Therefore, from (6.16)

$$\sum_{k=0}^{n}\binom{n}{k}(-1)^{k-1}x^k = (-1)^{n-1}(x-1)^n = -(1-x)^n \tag{6.17}$$

which matches the binomial formula.

6.3 Identities with Special Numbers

This section contains some new identities for products of harmonic, Bernoulli and Fibonacci numbers. We start with the following lemma.

Lemma 6.7 *For any two integers $1 \leq m \leq n$, we have*

$$\sum_{k=m}^{n}\binom{n}{k}\binom{k}{m}\frac{(-1)^k}{k} = \frac{(-1)^m}{m}. \tag{6.18}$$

The proof is given in Section 6.5.

Example 6.7 Let $a_k = \dfrac{(-1)^k}{k}$, $k \geq 1$. Then for any two integers $1 \leq m \leq n$ we have from the above lemma and from Lemma 6.2

$$\binom{n}{m}\nabla^m b_n = \sum_{k=1}^{n}\binom{n}{k}\binom{k}{m}a_k = \frac{(-1)^m}{m}. \tag{6.19}$$

Then from Theorem 6.1

$$\sum_{k=1}^{n}\binom{n}{k}\frac{(-1)^{k}c_{k}}{k} = \sum_{m=1}^{n}\frac{(-1)^{m}d_{m}}{m} . \qquad (6.20)$$

This is property (5.6) of the binomial transform. It is true for any two sequences $\{c_k\},\{d_k\}, k \geq 1$ related by (6.3). Note that (6.3) can be written in the form

$$\sum_{k=0}^{n}\binom{n}{k}(-1)^{k}c_{k} = (-1)^{n}d_{n} .$$

Example 6.8   This example uses the previous one. Let

$$H_{n} = 1+\frac{1}{2}+...+\frac{1}{n},\ H_{0} = 0,\ n \geq 0$$

be the harmonic numbers. Then we have (equation (3.22))

$$\sum_{k=1}^{n}\binom{n}{k}\frac{(-1)^{k-1}}{k} = H_{n} . \qquad (6.21)$$

According to Lemma 6.2 we find

$$\sum_{k=j}^{n}\binom{n}{k}\binom{k}{m}\frac{(-1)^{k}}{k} = \binom{n}{m}\nabla^{m}H_{n}$$

and then from Lemma 6.7 for $1 \leq m \leq n$

$$\binom{n}{m}\nabla^{m}H_{n} = \frac{(-1)^{m}}{m} . \qquad (6.22)$$

Therefore, from (6.21) and Corollary 6.6,

$$\sum_{k=1}^{n}\binom{n}{k}\frac{(-1)^{k-1}x^{k}}{k} = H_{n}-\sum_{k=1}^{n}\frac{(1-x)^{k}}{k} . \qquad (6.23)$$

This identity can easily be verified by differentiation.

Example 6.9   Formula (6.16) can also be used to evaluate the iterated differences $\nabla^m b_n$ when the left hand side there is known. By inversion in (6.21) we have

$$\sum_{k=0}^{n}\binom{n}{k}(-1)^{k-1}H_k = \frac{1}{n}. \tag{6.24}$$

The version with x is now (see (3.32))

$$\sum_{k=0}^{n}\binom{n}{k}(-1)^{k-1}H_k x^k$$

$$= \frac{1}{n}+\frac{1-x}{n-1}+\frac{(1-x)^2}{n-2}+...+\frac{(1-x)^{n-2}}{2}+\frac{(1-x)^{n-1}}{1}-(1-x)^n H_n .$$

Comparing this to (6.16) we conclude that

$$\binom{n}{m}\nabla^m \frac{1}{n}=\frac{(-1)^m}{n-m} \tag{6.25}$$

when $0 \le m < n$ and

$$\nabla^n \frac{1}{n} = (-1)^{n-1}H_n.$$

From Example 6.9 and Theorem 6.1 we derive the following interesting result.

Corollary 6.8 *Let $\{c_k\}$ and $\{d_k\}$ be any two sequences related as in (6.3). Then*

$$\sum_{k=0}^{n}\binom{n}{k}(-1)^{k-1}H_k c_k = (-1)^{n-1}H_n d_n + \sum_{m=0}^{n-1}\frac{(-1)^m d_m}{n-m}. \tag{6.26}$$

For the proof we take $a_k = (-1)^{k-1} H_k$ in Theorem 6.1 and then in view of (6.24) we have $b_n = \dfrac{1}{n}$. The rest follows from (6.25).

To show this corollary in action we shall give several examples.

Example 6.10 Applying property (5.6) to equation (6.24) we find

$$\sum_{k=1}^{n} \binom{n}{k} (-1)^{k-1} \frac{H_k}{k} = 1 + \frac{1}{2^2} + \dots + \frac{1}{n^2} .$$

With the notation

$$H_n^{(2)} = 1 + \frac{1}{2^2} + \dots + \frac{1}{n^2}, \quad H_0^{(2)} = 0$$

we obtain by inversion $(n \geq 1)$

$$\frac{H_n}{n} = \sum_{k=0}^{n} \binom{n}{k} (-1)^{k-1} H_k^{(2)}$$

and (6.26) yields (with $c_k = \dfrac{H_k}{k}$, $d_m = (-1)^{m-1} H_m^{(2)}$)

$$\sum_{k=0}^{n} \binom{n}{k} (-1)^{k-1} \frac{H_k^2}{k} = H_n H_n^{(2)} - \sum_{m=0}^{n-1} \frac{H_m^{(2)}}{n-m} . \qquad (6.27)$$

Example 6.11 In the same way, starting from the identity (see (5.64))

$$\frac{H_n}{n+1} = \sum_{k=0}^{n} \binom{n}{k} (-1)^{k-1} \frac{H_k}{k+1}$$

and taking $c_k = \dfrac{H_k}{k+1}$, $d_m = \dfrac{(-1)^{m-1} H_m}{m+1}$ in (6.26) we obtain

$$\sum_{k=0}^{n}\binom{n}{k}(-1)^{k-1}\frac{H_k^2}{k+1} = \frac{H_n^2}{n+1} - \sum_{m=0}^{n-1}\frac{H_m}{(n-m)(m+1)} . \qquad (6.28)$$

Now we shall prove several identities involving other special numbers.

Example 6.12 Here we give a new proof of (5.35), now with generalized Stieltjes numbers $S(\alpha, k)$, $\operatorname{Re}\alpha > 0$. By inversion in (2.1)

$$n^\alpha = \sum_{k=0}^{n}\binom{n}{k}k!S(\alpha, k)$$

and from (6.26) with $c_k = k^\alpha$ and $d_k = k!S(\alpha, k)$

$$\sum_{k=0}^{n}\binom{n}{k}(-1)^{k-1}H_k k^\alpha \qquad (6.29)$$

$$= (-1)^{n-1}n!H_n S(\alpha, n) + \sum_{m=0}^{n-1}\frac{(-1)^m m!S(\alpha, m)}{n-m} .$$

Example 6.13 For the Fibonacci numbers F_n we have (Example 5.7)

$$F_n = \sum_{k=0}^{n}\binom{n}{k}(-1)^{k-1}F_k, \quad F_{2n} = \sum_{k=0}^{n}\binom{n}{k}F_k .$$

From this and (6.26) we derive two new identities involving products of harmonic and Fibonacci numbers. Namely,

$$\sum_{k=0}^{n}\binom{n}{k}(-1)^{k-1}H_k F_k = H_n F_n - \sum_{m=0}^{n-1}\frac{F_m}{n-m} \qquad (6.30)$$

$$\sum_{k=0}^{n}\binom{n}{k}(-1)^{k-1}H_k F_{2k} = (-1)^{n-1}H_n F_n + \sum_{m=0}^{n-1}\frac{(-1)^m F_m}{n-m} . \qquad (6.31)$$

Example 6.14 Here we use the Bernoulli numbers B_n defined by the generating function

$$\frac{t}{e^t - 1} = \sum_{n=0}^{\infty} B_n \frac{t^n}{n!}, \quad |t| < 2\pi .$$

For the Bernoulli numbers it is known that (see Chapter 3, Example 3.1)

$$(-1)^n B_n = \sum_{k=0}^{n} \binom{n}{k} B_k .$$ (6.32)

From (6.26) with $c_k = (-1)^k B_k$, $d_m = B_m$ we obtain the identity

$$\sum_{k=0}^{n} \binom{n}{k} H_k B_n = (-1)^n H_n B_n - \sum_{m=0}^{n-1} \frac{(-1)^m B_m}{n-m} .$$ (6.33)

Example 6.15 In this example we use the Laguerre polynomials

$$L_n(x) = \frac{e^x}{n!} \left(\frac{d}{dx} \right)^n (x^n e^{-x})$$

which satisfy the identity

$$L_n(x) = \sum_{k=0}^{n} \binom{n}{k} \frac{(-x)^k}{k!} .$$

Now (6.26) provides the curious formula

$$\sum_{k=0}^{n} \binom{n}{k} (-1)^k H_k L_k(x) = \frac{x^n}{n!} H_n - \sum_{m=0}^{n-1} \frac{x^m}{m!(n-m)} .$$ (6.34)

Next we return to the sequence of Fibonacci numbers defined by the recurrence $F_n = F_{n-1} + F_{n-2}$ and starting from $F_0 = 0$, $F_1 = 1$. We can extend the sequence F_n to negative indices by using the equation $F_{n-2} = F_n - F_{n-1}$. Thus we come to the "negative" Fibonacci numbers, where $F_{-n} = (-1)^{n+1} F_n$, $n \geq 0$. Computing the backward differences we find

$$\nabla F_n = F_n - F_{n-1} = F_{n-2}$$

$$\nabla^2 F_n = F_{n-2} - F_{n-3} = F_{n-4}$$

etc. Obviously,

$$\nabla^m F_n = F_{n-2m}$$

and this is true for any non-negative integer m. Now we can prove the following result.

Corollary 6.9 *For any pair of sequences $\{c_k\}$ and $\{d_k\}$ as in (6.3) and for every non-negative integer n we have*

$$\sum_{k=0}^{n} \binom{n}{k} (-1)^{k-1} F_k\, c_k = \sum_{m=0}^{n} \binom{n}{m} d_m F_{n-2m} . \qquad (6.35)$$

For the proof we apply equation (6.4) from Theorem 6.1 with $a_k = (-1)^{k-1} F_k$ and $b_n = F_n$.

Formula (6.35) can be used in the same way as (6.26) to generate various new identifies with Fibonacci numbers by choosing different sequences $\{c_k\}$. For illustration we provide the following example:

Example 6.16 Choosing $c_n = (-1)^n B_n$ and $d_k = B_k$, where B_n are the Bernoulli numbers, we obtain from (6.32) and (6.35) the following identity connecting Bernoulli and Fibonacci numbers

$$\sum_{k=0}^{n} \binom{n}{k} B_k F_k = -\sum_{m=0}^{n} \binom{n}{m} B_m F_{n-2m} . \qquad (6.36)$$

The Lucas numbers L_n satisfy the same recurrence $L_n = L_{n-1} + L_{n-2}$ as the Fibonacci numbers and for them we have the similar binomial identities (5.52) and (5.53). Therefore, a property similar to (6.35) is also true for the Lucas numbers. We leave this to the reader.

6.4   Some Variations

Remark 6.10   *If the binomial transform is defined by the formula*

$$\sum_{k=0}^{n} \binom{n}{k} (-1)^k a_k = b_n \tag{6.37}$$

then (6.16) takes the form

$$\sum_{k=0}^{n} \binom{n}{k} a_k (-x)^k = \sum_{m=0}^{n} \binom{n}{m} \nabla^m b_n (x-1)^m$$

or

$$\sum_{k=0}^{n} \binom{n}{k} a_k x^k = \sum_{m=0}^{n} \binom{n}{m} (-1)^m \nabla^m b_n (x+1)^m . \tag{6.38}$$

Remark 6.11   *The coefficients appearing in Lemma 6.2*

$$C(n,m) = \binom{n}{m} \nabla^m b_n$$

can be written in terms of Stirling numbers of the first kind $s(m,j)$. Suppose the sequence $\{b_n\}$ is the binomial transform of the sequence $\{a_n\}$ as in (6.1). Then

$$C(n,m) = \frac{1}{m!} \sum_{j=0}^{m} s(m,j) (n\nabla)^j b_n . \tag{6.39}$$

Proof. From (2.5) we have the representation

$$\binom{k}{m} = \frac{1}{m!}\sum_{j=0}^{m} s(m, j)k^{j}$$

and from this and (6.7)

$$C(n,m) = \sum_{k=0}^{n}\binom{n}{k}\binom{k}{m}a_{k} = \frac{1}{m!}\sum_{k=0}^{n}\binom{n}{k}\left\{\sum_{j=0}^{m} s(m, j)k^{j}a_{k}\right\}$$

$$= \frac{1}{m!}\sum_{j=0}^{m} s(m, j)\left\{\sum_{k=0}^{n}\binom{n}{k}k^{j}a_{k}\right\}.$$

Therefore, in view of Theorem 5.1 we come to (6.39).   □

6.5   Proofs

Proof of Lemma 6.2.   When $m=0$ or $m=1$ the lemma is true for every $n \geq 0$, as (6.6) becomes (6.8) for $m=1$. We shall do induction on $1 \leq m \leq n$. Suppose the identity is true for some $m < n$. We shall prove it for $m+1$.

The left hand side in (6.6) becomes (using (6.8) in the second equality)

$$\sum_{k=0}^{n}\binom{n}{k}k(k-1)...(k-m+1)(k-m)a_{k}$$

$$= \sum_{k=0}^{n}\binom{n}{k}k\left\{k(k-1)...(k-m+1)a_{k}\right\} - m\sum_{k=0}^{n}\binom{n}{k}k(k-1)...(k-m+1)a_{k}$$

$$= n\nabla\left\{m!\binom{n}{m}\nabla^{m}b_{n}\right\} - m\left\{m!\binom{n}{m}\nabla^{m}b_{n}\right\}$$

$$= m! \left\{ n \binom{n}{m} \nabla^m b_n - n \binom{n-1}{m} \nabla^m b_{n-1} - m \binom{n}{m} \nabla^m b_n \right\}$$

$$= m! \left\{ (n-m) \binom{n}{m} \nabla^m b_n - n \binom{n-1}{m} \nabla^m b_{n-1} \right\}$$

$$= m! \left\{ \frac{n!}{(n-m-1)!m!} \nabla^m b_n - \frac{n!}{(n-m-1)!m!} \nabla^m b_{n-1} \right\}$$

$$= m! \left\{ \frac{n!}{(n-m-1)!m!} \nabla^{m+1} b_n \right\} = (m+1)! \binom{n}{m+1} \nabla^{m+1} b_n . \qquad \square$$

Proof of Lemma 6.3. From Lemma 6.2 and the inversion formula for the binomial transform we have

$$\binom{n}{m} \nabla^m b_n = \frac{1}{m!} \sum_{k=0}^{n} \binom{n}{k} k(k-1)...(k-m+1) a_k$$

$$= \sum_{k=0}^{n} \binom{n}{k} \binom{k}{m} a_k = \sum_{k=0}^{n} \binom{n}{k} \binom{k}{m} \left\{ \sum_{j=0}^{k} \binom{k}{j} (-1)^{k-j} b_j \right\}$$

$$= \sum_{j=0}^{n} (-1)^j b_j \left\{ \sum_{k=0}^{n} \binom{n}{k} \binom{k}{m} \binom{k}{j} (-1)^k \right\}$$

$$= \sum_{j=0}^{n} \binom{n}{j} \binom{j}{n-m} (-1)^{n-j} b_j$$

where in the last equality we are using the identity

$$\sum_{k=0}^{n} \binom{n}{k} \binom{k}{m} \binom{k}{j} (-1)^k = (-1)^n \binom{n}{j} \binom{j}{n-m}$$

(see [59], p. 15). $\qquad \square$

Proof of Theorem 6.1.

$$\sum_{k=0}^{n}\binom{n}{k}a_k c_k = \sum_{k=0}^{n}\binom{n}{k}a_k\left\{\sum_{k=0}^{n}\binom{k}{m}d_m\right\}$$

$$= \sum_{m=0}^{n}d_m\left\{\sum_{k=0}^{n}\binom{n}{k}\binom{k}{m}a_k\right\} = \sum_{m=0}^{n}\binom{n}{m}d_m\nabla^m b_n$$

according to Lemma 6.2.   □

Proof of Lemma 6.7.   The starting point of this proof is the identity (6.12) with $1\le j\le n$. We divide both sides of this identity by x and integrate from 0 to 1. This yields

$$\sum_{k=j}^{n}\binom{n}{k}\binom{k}{j}\frac{(-1)^k}{k} = (-1)^j\binom{n}{j}\int_0^1 x^{j-1}(1-x)^{n-j}dx$$

$$= (-1)^j\binom{n}{j}B(n-j+1,j) = \frac{(-1)^j}{j} .$$

The evaluation of the integral is from the table [36], namely, this is entry 3.191 (3). Here $B(x,y)$ is Euler's Beta function,

$$B(x,y) = \frac{\Gamma(x)\Gamma(y)}{\Gamma(x+y)}$$

so that

$$B(n-j+1,j) = \frac{(n-j)!(j-1)!}{n!} = \frac{(n-j)!j!}{jn!} = \frac{1}{j}\binom{n}{j}^{-1}.$$

The proof is completed.   □

Chapter 7

Special Formulas and Power Series with Binomial Sums

In this chapter we give a collection of various formulas involving binomial expressions for easy reference. The formulas can be used for computing binomial transforms or for other purposes.

In certain cases, the binomial transform can be evaluated by using the Nörlund-Rice integral formula (Flajolet and Sedgewick [29] and also Nörlund [54])

$$\sum_{k=0}^{n} \binom{n}{k} (-1)^k f(k) = \frac{(-1)^n n!}{2\pi i} \oint_L \frac{f(z)\, dz}{z(z-1)(z-2)...(z-n)} \quad (7.1)$$

where $f(z)$ is analytic in a domain containing the interval $[0,n]$ and L is a closed, simple, positively oriented curve surrounding the interval $[0,n]$. For some applications of this formula see also [41] and [42].

Using the difference $\Delta f(x) = f(x+1) - f(x)$ we have by iteration

$$\sum_{k=0}^{n} \binom{n}{k} (-1)^k f(x+k) = (-1)^n \Delta^n f(x) \quad (7.2)$$

and by inversion,

$$f(x+n) = \sum_{k=0}^{n} \binom{n}{k} \Delta^k f(x)$$

(Graham et al., [37, (5.40)]). In particular,

$$\sum_{k=0}^{n}\binom{n}{k}(-1)^{k}f(k) = (-1)^{n}\Delta^{n}f(0). \tag{7.3}$$

Given a polynomial or a formal power series

$$f(t) = a_{0} + a_{1}t + a_{2}t^{2} + ... \tag{7.4}$$

we have the following theorem (see [11]).

Theorem 7.1 *For any power series $f(t)$ as in (7.4) and for any integer $n \geq 0$*

$$\sum_{k=0}^{n}\binom{n}{k}(-1)^{k}f(y+zk) \tag{7.5}$$

$$= (-1)^{n}n!\sum_{m=0}^{\infty}a_{m}\left\{\sum_{p=0}^{m}S(p,n)z^{p}y^{m-p}\right\}$$

where y, z are parameters and $S(p,n)$ are the Stirling numbers of the second kind. In particular, for $y = 0$,

$$\sum_{k=0}^{n}\binom{n}{k}(-1)^{k}f(zk) = (-1)^{n}n!\sum_{m=0}^{\infty}a_{m}z^{m}S(m,n). \tag{7.6}$$

If $f(t)$ is analytic on the disk $|t| \leq R$, then for every $n > 0$ the power series on the right hand side in (7.6) converges in the disk $|z| < R / n$.

Proof. Starting from the binomial expansion

$$(y+zk)^{m} = \sum_{p=0}^{m}\binom{m}{p}z^{p}k^{p}y^{m-p}$$

we write by changing the order of summation

$$\sum_{k=0}^{n}\binom{n}{k}(-1)^{k}(y+zk)^{m} = \sum_{k=0}^{n}\binom{n}{k}(-1)^{k}\left\{\sum_{p=0}^{m}\binom{m}{p}z^{p}k^{p}y^{m-p}\right\}$$

$$= \sum_{p=0}^{m} \binom{m}{p} z^{p} y^{m-p} \left\{ \sum_{k=0}^{n} \binom{n}{k} (-1)^{k} k^{p} \right\}$$

$$= (-1)^{n} n! \sum_{p=0}^{m} \binom{m}{p} z^{p} y^{m-p} S(p,n) \ .$$

That is,

$$\sum_{k=0}^{n} \binom{n}{k} (-1)^{k} (y+zk)^{m} \qquad (7.7)$$

$$= (-1)^{n} n! \sum_{p=0}^{m} \binom{m}{p} S(p,n) z^{p} y^{m-p} \ .$$

Multiplying both sides by a_m and summing for $m = 0,1,2...$ we obtain (7.5).

Now we prove convergence in (7.6). Let L be the circle $|\lambda| = R$. For every $m \geq 0$ we have by Cauchy's integral formula

$$a_{m} = \frac{1}{2\pi i} \oint_{L} \frac{f(\lambda)d\lambda}{\lambda^{m+1}} \ .$$

Suppose now $|z| < R/n$. Using (2.2) and changing the order of summation and integration we have

$$\sum_{m=0}^{\infty} S(m,n) a_{m} z^{m} = \frac{1}{2\pi i} \oint_{L} \left\{ \sum_{m=0}^{\infty} S(m,n) \left(\frac{z}{\lambda} \right)^{m} \right\} \frac{f(\lambda)d\lambda}{\lambda}$$

$$= \frac{z^{n}}{2\pi i} \oint_{L} \left\{ \left(1 - \frac{z}{\lambda} \right) \left(1 - \frac{2z}{\lambda} \right) \cdots \left(1 - \frac{nz}{\lambda} \right) \right\}^{-1} \frac{f(\lambda)d\lambda}{\lambda^{n+1}}$$

$$= \frac{z^{n}}{2\pi i} \oint_{L} \left\{ (\lambda - z)(\lambda - 2z) \cdots (\lambda - nz) \right\}^{-1} \frac{f(\lambda)d\lambda}{\lambda}$$

$$= \frac{1}{2\pi i} \oint_L \left\{ \left(\frac{\lambda}{z} - 1 \right)\left(\frac{\lambda}{z} - 2 \right) \cdots \left(\frac{\lambda}{z} - n \right) \right\}^{-1} \frac{f(\lambda)d\lambda}{\lambda} \; .$$

With the substitution $\lambda = \mu z$, this becomes

$$= \frac{1}{2\pi i} \oint_M \frac{f(\mu z)d\mu}{\mu(\mu - 1)(\mu - 2)\cdots(\mu - n)}$$

where M is now a circle centered at the origin and containing the numbers $1, 2, .., n$. According to the Nörlund-Rice formula (7.1) we have

$$\frac{1}{2\pi i} \oint_M \frac{f(\mu z)d\mu}{\mu(\mu - 1)(\mu - 2)\cdots(\mu - n)} = \frac{(-1)^n}{n!} \sum_{k=0}^{n} \binom{n}{k}(-1)^k f(kz)$$

and the proof is completed.   □

Remark 7.2   *The summation in the series on the right hand side in (7.5) and (7.6) starts, in fact, from $m = n$, since $S(m,n) = 0$ when $m < n$. It follows from (7.5) that for any y, z,*

$$\sum_{k=0}^{n} \binom{n}{k}(-1)^k (y + zk)^m = 0 \quad (m < n) \tag{7.8}$$

and the sum equals $(-1)^n n! z^n$ for $m = n$ (see [2]).

In order to show one interesting application of the above theorem we need a simple lemma (see also equation (17) on p. 189 in [39]).

Lemma 7.3   *For every integer $m > 1$ we have*

$$\sum_{n=1}^{m} S(m,n)(n-1)! \, (-1)^n = 0 \tag{7.9}$$

and when $m = 1$ the value of this sum is -1.

Proof. The Stirling numbers of the second kind satisfy the recurrence relation

$$S(m,n) = n\,S(m-1,n) + S(m-1,n-1), \quad 1 \le n \le m$$

and the lemma follows from here by a simple computation. Let $m \ge 2$. Then

$$\sum_{n=1}^{m} S(m,n)(n-1)!\,(-1)^n$$

$$= \sum_{n=1}^{m} S(m-1,n)n!\,(-1)^n + \sum_{n=1}^{m} S(m-1,n-1)(n-1)!\,(-1)^n$$

$$= \sum_{n=1}^{m-1} S(m-1,n)n!\,(-1)^n + \sum_{k=0}^{m-1} S(m-1,k)k!\,(-1)^{k+1}$$

(by setting $n-1=k$ and noticing that $S(m-1,m)=0$)

$$= \sum_{n=1}^{m-1} S(m-1,n)n!\,(-1)^n - \sum_{k=1}^{m-1} S(m-1,k)k!\,(-1)^k - S(m-1,0) = 0.$$

The case $m=1$ is obvious and the proof is completed.   □

From the above lemma and Theorem 7.1 we derive the corollary:

Corollary 7.4   *Let* $f(t)$ *be a formal power series as in (7.4). Then for every* z *we have*

$$\sum_{n=1}^{\infty} \frac{1}{n} \left\{ \sum_{k=0}^{n} \binom{n}{k}(-1)^k f(zk) \right\} = -f'(0)z. \tag{7.10}$$

Proof. We multiply both sides in equation (7.6) by $\dfrac{1}{n}$ and sum for $n = 1,2,\dots$. Changing the order of summation yields

$$\sum_{n=1}^{\infty} \frac{1}{n} \left\{ \sum_{k=0}^{n} \binom{n}{k}(-1)^k f(zk) \right\}$$

$$= \sum_{m=0}^{\infty} a_m z^m \left\{ \sum_{n=1}^{m} S(m,n)(n-1)!(-1)^n \right\} = -a_1 z$$

which is the desired result.   □

When $f(t)$ is a polynomial of degree p the series on the left hand side in (7.10) truncates, as the binomial sums become zeros for $n > p$ (see Remark 7.2).

A large class of functions for which (7.10) holds is presented in [14] together with several applications.

We continue with results related to Theorem 7.1.

Example 7.1   Consider the polynomial

$$g(t) = \binom{t}{p} = \frac{1}{p!} \sum_{m=0}^{p} s(p,m) t^m \tag{7.11}$$

where $p \geq 0$ is a nonnegative integer and $s(m,p)$ are the Stirling numbers of the first kind — see (2.4) and (2.5). Clearly, $\deg(g) = p$. Applying equation (7.6) we find the identity

$$\sum_{k=0}^{n} \binom{n}{k}(-1)^k \binom{zk}{p} = (-1)^n \frac{n!}{p!} \sum_{m=0}^{p} s(p,m) S(m,n) z^m. \tag{7.12}$$

These sums appeared in Todorov's paper [71] in the Taylor series expansion

$$\left((1+t)^z - 1 \right)^n = \sum_{p=0}^{\infty} t^p \left\{ (-1)^n \sum_{k=0}^{n} \binom{n}{k}(-1)^k \binom{zk}{p} \right\}.$$

When $p < n$, both sides in (7.12) are zeros.

Now we want to mention two series transformation theorems involving binomial transforms. First, consider the exponential polynomials (2.22)

$$\varphi_m(x) = \sum_{n=0}^{m} S(m,n)x^n .$$

These polynomials appear in the right hand side of the expansion given in the following theorem.

Theorem 7.5   *With $f(t)$ as in (7.4) and with parameters y, z, we have the series transformation*

$$\sum_{n=0}^{\infty} \frac{x^n}{n!} \left\{ \sum_{k=0}^{n} \binom{n}{k} (-1)^k f(y+zk) \right\} \tag{7.13}$$

$$= \sum_{m=0}^{\infty} a_m \left\{ \sum_{p=0}^{m} \binom{m}{p} z^p y^{m-p} \varphi_p(-x) \right\}.$$

When $y = 0$ this becomes

$$\sum_{n=0}^{\infty} \frac{x^n}{n!} \left\{ \sum_{k=0}^{n} \binom{n}{k} (-1)^k f(zk) \right\} = \sum_{m=0}^{\infty} a_m \varphi_m(-x) z^m . \tag{7.14}$$

Proof. Multiplying both sides of (7.5) by $x^n / n!$ and summing for $n = 0, 1...$, we come to (7.13) after changing the order of summation on the right hand side. Note that since $S(p,n) = 0$ for $n > p$, the summation on the right hand side leads exactly to the exponential polynomials.   □

We do not specify convergence in the above formulas (and also in many of those below). If $f(t)$ is an entire function, the partial sums of the series on the left hand side in (7.13) are defined for every z. If the series on the right hand side converges for some number $z = a$, then the series will converge absolutely in the disk $|z| < a$ being a power series. Depending on the choice of the function $f(t)$ the series on the right hand side may diverge. It was shown in [11] that in certain cases such series represent asymptotic expansions.

From the above theorem with $x = -1$, we deduce the immediate corollary.

Corollary 7.6   *For any power series $f(t)$ as in (7.4) we have*

$$\sum_{n=0}^{\infty} \frac{(-1)^n}{n!} \left\{ \sum_{k=0}^{n} \binom{n}{k} (-1)^k f(zk) \right\} = \sum_{m=0}^{\infty} a_m b_m z^m \qquad (7.15)$$

where the numbers

$$b_n = \varphi_n(1) = \sum_{k=0}^{n} S(n,k)$$

are the Bell numbers [5], [6].

Example 7.2   Using Theorem 7.5 and (7.12) we obtain the representation

$$\sum_{n=0}^{p} \frac{x^n}{n!} \left\{ \sum_{k=0}^{n} \binom{n}{k} (-1)^k \binom{zk}{p} \right\} = \frac{1}{p!} \sum_{m=0}^{p} s(p,m) \, \varphi_m(-x) \, z^m \qquad (7.16)$$

for every integer $p \geq 0$.

We shall find now analogous expansions involving the geometric polynomials (2.27)

$$\omega_m(x) = \sum_{n=0}^{m} S(m,n) n! x^n.$$

Multiplying both sides of (7.5) by x^n and summing for $n = 0, 1, \ldots$, we obtain (after changing the order of summation) the following analogue to Theorem 7.6.

Theorem 7.7   *With $f(t)$ as in (7.4) and with parameters y, z*

$$\sum_{n=0}^{\infty} x^n \left\{ \sum_{k=0}^{n} \binom{n}{k} (-1)^k f(y + zk) \right\} \qquad (7.17)$$

$$= \sum_{m=0}^{\infty} a_m \left\{ \sum_{p=0}^{m} \binom{m}{p} z^p y^{m-p} \, \omega_p(-x) \right\} .$$

When $y = 0$, this becomes

$$\sum_{n=0}^{\infty} x^n \left\{ \sum_{k=0}^{n} \binom{n}{k} (-1)^k f(zk) \right\} = \sum_{m=0}^{\infty} a_m z^m \, \omega_m(-x) . \qquad (7.18)$$

Example 7.3   From (7.11) and (7.18) we have for all integers $p \geq 0$

$$\sum_{n=0}^{p} x^n \left\{ \sum_{k=0}^{n} \binom{n}{k} (-1)^k \binom{zk}{p} \right\} = \frac{1}{p!} \sum_{m=0}^{p} s(p,m) \, \omega_m(-x) z^m . \qquad (7.19)$$

Example 7.4   Integration for x in (7.17) yields

$$\sum_{n=0}^{\infty} \frac{x^{n+1}}{n+1} \left\{ \sum_{k=0}^{n} \binom{n}{k} (-1)^k f(y+zk) \right\}$$

$$= \sum_{m=0}^{\infty} a_m \left\{ \sum_{p=0}^{m} \binom{m}{p} z^p y^{m-p} \int_0^x \omega_p(-t) \, dt \right\} .$$

After evaluating the integral on the right hand side and reducing by x

$$\sum_{n=0}^{\infty} \frac{x^n}{n+1} \left\{ \sum_{k=0}^{n} \binom{n}{k} (-1)^k f(y+zk) \right\} \qquad (7.20)$$

$$= \sum_{m=0}^{\infty} a_m \left\{ \sum_{p=0}^{m} \binom{m}{p} z^p y^{m-p} \sum_{j=0}^{p} S(p,j) j! \, \frac{(-x)^j}{j+1} \right\} .$$

Setting here $x = 1, y = 0$, we find

$$\sum_{n=0}^{\infty} \frac{1}{n+1} \left\{ \sum_{k=0}^{n} \binom{n}{k} (-1)^k f(zk) \right\} = \sum_{m=0}^{\infty} a_m z^m \sum_{j=0}^{m} S(m,j) \, j! \frac{(-1)^j}{j+1} \, .$$

Using the well-known representation of the Bernoulli numbers in terms of Stirling numbers of the second kind

$$B_n = \sum_{j=0}^{m} S(m,j) j! \frac{(-1)^j}{j+1}$$

we can write now the interesting representation

$$\sum_{n=0}^{\infty} \frac{1}{n+1} \left\{ \sum_{k=0}^{n} \binom{n}{k} (-1)^k f(zk) \right\} = \sum_{m=0}^{\infty} a_m B_m z^m \, . \qquad (7.21)$$

More results in this direction can be found in [11].

Part 2

Table of Binomial Transforms

Chapter 8

Assorted Binomial Formulas

(8.1)
$$\sum_{k=0}^{n}\binom{n}{k}x^{k}=(x+1)^{n}.$$

Multiplying this by x^{m-1} and integrating (using the substitution $u=x+1$) we obtain

(8.2)
$$\sum_{k=0}^{n}\binom{n}{k}\frac{x^{m+k}}{m+k}=\sum_{j=1}^{m}(-1)^{m-j}\binom{m-1}{m-j}\frac{(x+1)^{n+j}-1}{n+j}.$$

In particular,

(8.3)
$$\sum_{k=0}^{n}\binom{n}{k}\frac{x^{k+1}}{k+1}=\frac{(x+1)^{n+1}-1}{n+1}$$

(8.4)
$$\sum_{k=0}^{n}\binom{n}{k}\frac{1}{k+1}=\frac{2^{n+1}-1}{n+1}.$$

With $x=-1, n>0$ in (8.1)

(8.5)
$$\sum_{k=0}^{n}\binom{n}{k}(-1)^{k}=0.$$

With summation starting from $k=1$,

$$(8.6) \qquad \sum_{k=1}^{n} \binom{n}{k}(-1)^{k-1} = 1$$

and with $x=1$ in (8.1),

$$(8.7) \qquad \sum_{k=0}^{n} \binom{n}{k} = 2^n .$$

By consecutive differentiation in (8.1),

$$(8.8) \qquad \sum_{k=0}^{n} \binom{n}{k}k = n2^{n-1}$$

$$(8.9) \qquad \sum_{k=0}^{n} \binom{n}{k}(-1)^k k = 0 \quad (n>1)$$

$$(8.10) \qquad \sum_{k=0}^{n} \binom{n}{k}k^2 = n(n+1)2^{n-2}$$

$$(8.11) \qquad \sum_{k=0}^{n} \binom{n}{k}k^3 = n^2(n+3)2^{n-3}$$

$$(8.12) \qquad \sum_{k=0}^{n} \binom{n}{k}k^4 = n(n+1)(n^2+5n-2)2^{n-4}$$

$$(8.13) \qquad \sum_{k=0}^{n} \binom{n}{k}k^5 = (n^5+10n^4+15n^3-10n^2)2^{n-5}$$

$$(8.14) \qquad \sum_{k=0}^{n} \binom{n}{k}k(k-1) = n(n-1)2^{n-2}$$

$$(8.15) \qquad \sum_{k=0}^{n} \binom{n}{k}x^k(x+2)^k = (x+1)^{2n} = [1+x(x+2)]^n .$$

For any complex α with $\operatorname{Re}\alpha > 0$ (see (6.14) and also [3])

$$(8.16) \qquad \sum_{k=0}^{n} \binom{n}{k} k^{\alpha} x^{k} = \sum_{j=0}^{n} \binom{n}{j} j! S(\alpha, j) \, x^{j} (1+x)^{n-j}$$

$$= (1+x)^{n} \sum_{j=0}^{n} \binom{n}{j} j! S(\alpha, j) \left(\frac{x}{1+x}\right)^{j}$$

or, changing x to $-x$

$$(8.17) \qquad \sum_{k=0}^{n} \binom{n}{k} (-1)^{k} k^{\alpha} x^{k}$$

$$= \sum_{j=0}^{n} \binom{n}{j} (-1)^{j} j! S(\alpha, j) \, x^{j} (1-x)^{n-j}$$

$$= (1-x)^{n} \sum_{j=0}^{n} \binom{n}{j} (-1)^{j} j! S(\alpha, j) \left(\frac{x}{1-x}\right)^{j}$$

where $S(\alpha, k)$ are the Stirling numbers of the second kind — see (2.1). With $x = 1$ in (8.16),

$$(8.18) \qquad \sum_{k=0}^{n} \binom{n}{k} k^{\alpha} = \sum_{j=0}^{n} \binom{n}{j} j! S(\alpha, j) \, 2^{n-j}$$

and with $x = -1$ in (8.16) we come to (2.1), namely,

$$(8.19) \qquad \sum_{k=0}^{n} \binom{n}{k} (-1)^{k} k^{\alpha} = (-1)^{n} n! S(\alpha, n) \ .$$

For any positive integer m and with parameters y, z

$$(8.20) \qquad \sum_{k=0}^{n} \binom{n}{k} (-1)^{k} (y + zk)^{m}$$

$$= (-1)^n n! \sum_{p=0}^{m} \binom{m}{p} S(p,n) z^p y^{m-p} ,$$

where both sides are zero for $n > m$ (see (7.7) in Chapter 7). Inverting (8.19) we find

(8.21) $$\sum_{k=0}^{n} \binom{n}{k} k! S(\alpha,k) = \sum_{k=1}^{n} \binom{n}{k} k! S(\alpha,k) = n^\alpha .$$

From this, by using property (5.6), and changing α to $\alpha+1$ we find

(8.22) $$\sum_{k=1}^{n} \binom{n}{k} (k-1)! S(\alpha+1,k) = \sum_{k=1}^{n} k^\alpha$$

for any complex number α with $\operatorname{Re}\alpha > -1$. From here, by inversion,

$$\sum_{k=1}^{n} \binom{n}{k} (-1)^{n-k} \sum_{j=1}^{k} j^\alpha = (n-1)! S(\alpha+1,n)$$

or

(8.23) $$\sum_{k=1}^{n} \binom{n}{k} (-1)^k \sum_{j=1}^{k} j^\alpha = (-1)^n (n-1)! S(\alpha+1,n) .$$

Let m be a nonnegative integer. For the sequence of Stirling numbers $S(k,m)$, $k = 0,1,2,...$ we have

(8.24) $$\sum_{k=0}^{n} \binom{n}{k} S(k,m) = S(n+1,m+1)$$

(Graham *et al.*, [37], (615)).

Let $P_m(t) = a_0 + a_1 t + ... + a_m t^m$ be a polynomial of degree m. Similar to (8.20) we have

(8.25) $$\sum_{k=0}^{n}\binom{n}{k}(-1)^{k}P_{m}(k)=(-1)^{n}n!\sum_{j=0}^{m}a_{j}S(j,n)\ .$$

Generally, if $f(t)=a_{0}+a_{1}t+...$ is a formal power series (see Chapter 7)

(8.26) $$\sum_{k=0}^{n}\binom{n}{k}(-1)^{k}f(k)=(-1)^{n}n!\sum_{j=0}^{\infty}a_{j}S(j,n)\ .$$

For any polynomial $f(t)$, every number a, and every $\lambda\neq 0,1,2,...$

(8.27) $$\sum_{k=0}^{n}\binom{n}{k}(-1)^{k}\frac{f(a+k)}{k-\lambda}$$

$$=f(a+\lambda)\sum_{k=0}^{n}\binom{n}{k}\frac{(-1)^{k}}{k-\lambda}+(-1)^{n}n!\sum_{m=n+1}^{\infty}\frac{f^{(m)}(a)}{m!}\left\{\sum_{j=0}^{m-1}\lambda^{j}S(m-j-1,n)\right\}$$

(Melzak's formula. For details see Chapter 4.)
    With $f=1$ and $\lambda=-y$ in (8.27)

(8.28) $$\sum_{k=0}^{n}\binom{n}{k}(-1)^{k}\frac{1}{y+k}=\frac{n!}{y(y+1)(y+2)...(y+n)}$$

$$=\frac{1}{y}\binom{n+y}{n}^{-1}$$

equivalently

$$\sum_{k=0}^{n}\binom{n}{k}(-1)^{k}\frac{y}{y+k}=\binom{n+y}{n}^{-1}$$

and by inversion in the last equation

$$(8.29) \qquad \sum_{k=0}^{n} \binom{n}{k} (-1)^k \binom{k+y}{k}^{-1} = \frac{y}{y+k} .$$

When $y = 1$ in (8.28)

$$(8.30) \qquad \sum_{k=0}^{n} \binom{n}{k} (-1)^k \frac{1}{k+1} = \frac{1}{n+1}$$

or, with summation from $k = 1$

$$\sum_{k=1}^{n} \binom{n}{k} (-1)^{k-1} \frac{1}{k+1} = \frac{n}{n+1} .$$

This is the same as equation (3.23) obtained before by property (1.9). Applying now (1.10) to (8.30) we find for any integer $m > 0$

$$\sum_{k=0}^{n} \binom{n}{k} (-1)^{k+m-1} \frac{1}{k+m} = \Delta^{m-1} \frac{1}{n+1}$$

or

$$(8.31) \qquad \sum_{k=0}^{n} \binom{n}{k} (-1)^k \frac{1}{k+m} = (-1)^{m-1} \Delta^{m-1} \frac{1}{n+1}$$

(see (8.37) below for another form of the right hand side).

Applying (5.6) to (8.30) we find

$$(8.32) \qquad \sum_{k=1}^{n} \binom{n}{k} \frac{(-1)^{k-1}}{k(k+1)} = H_{n+1} - 1 .$$

Further,

$$(8.33) \qquad \sum_{k=0}^{n} \binom{n}{k} (-1)^k \frac{1}{(y+k)^2}$$

$$= \frac{n!}{y(y+1)(y+2)...(y+n)} \sum_{k=0}^{n} \frac{1}{k+y}$$

(8.34)
$$\sum_{k=0}^{n} \binom{n}{k} \frac{(-1)^k}{(y+k)(y+k+1)...(y+k+m)}$$

$$= \frac{(n+m)!}{m!\,y(y+1)...(y+n+m)}$$

(Chu and Yan [21]). With $m=0$ this reduces to (8.28). The dual is

(8.35)
$$\sum_{k=0}^{n} \binom{n}{k} (-1)^k \frac{(k+m)!}{y(y+1)...(y+k+m)}$$

$$= \frac{m!}{(y+n)(y+n+1)...(y+n+m)}$$

(8.36)
$$\sum_{k=0}^{n} \binom{n}{k} \frac{(-1)^k}{(k+m)^p}$$

$$= \frac{(m-1)!\,n!}{(m+n)!\,(p-1)!} Y_{p-1}(...,(i-1)!(H_{m+n}^{(i)} - H_{m-1}^{(i)}),...)$$

(Kirschenhofer [42], Larsen and Larcombre [47],). Here $Y_n(x_1, x_2, ... x_n)$ are the many variable Bell polynomials (Comtet, [22]), $m \ge 1, p \ge 1$, and $H_n^{(k)}$ are the extended harmonic numbers (2.6). In particular, we have (cf. (8.28))

(8.37)
$$\sum_{k=0}^{n} \binom{n}{k} \frac{(-1)^k}{k+m} = \frac{(m-1)!\,n!}{(m+n)!} = \frac{n!}{m(m+1)...(m+n)}$$

$$= \frac{1}{m} \binom{n+m}{n}^{-1}$$

(8.38) $$\sum_{k=0}^{n} \binom{n}{k} \frac{(-1)^k}{(k+m)^2} = \frac{(m-1)!n!}{(m+n)!}(H_{m+n} - H_{m-1})$$

(resulting from ((8.33) when $y=m$ is an integer). For $m=1, m=2$ and $m=3$ in (8.38) we have correspondingly

(8.39) $$\sum_{k=0}^{n} \binom{n}{k} \frac{(-1)^k}{(k+1)^2} = \frac{H_{n+1}}{n+1}$$

(8.40) $$\sum_{k=0}^{n} \binom{n}{k} \frac{(-1)^k}{(k+2)^2} = \frac{H_{n+2}-1}{(n+1)(n+2)}$$

(8.41) $$\sum_{k=0}^{n} \binom{n}{k} \frac{(-1)^k}{(k+3)^2} = \frac{2H_{n+3}-3}{(n+1)(n+2)(n+3)} \, .$$

In general

(8.42) $$\sum_{k=0}^{n} \binom{n}{k} \frac{(-1)^k}{(k+m)^3}$$

$$= \frac{(m-1)!n!}{2(m+n)!}[(H_{m+n} - H_{m-1})^2 + H_{m+n}^{(2)} - H_{m-1}^{(2)}] \, .$$

(8.43) $$\sum_{k=0}^{n} \binom{n}{k}(-1)^k \frac{1}{2k+1} = \frac{2.4....(2n)}{3.5....(2n+1)} = \frac{4^n}{2n+1} \binom{2n}{n}^{-1}$$

(see also its inverse (10.40)).

(8.44) $$\sum_{k=0}^{n} \binom{n}{k}(-1)^k \frac{1}{(2k+1)^2}$$

$$= \frac{4^n}{2n+1} \binom{2n}{n}^{-1} \left(1 + \frac{1}{3} + ... + \frac{1}{2n+1}\right)$$

(see Berndt [1], p. 293).

$$(8.45) \qquad \sum_{k=0}^{n} \binom{n}{k} (-1)^k \frac{1}{1-2k} = 4^n \binom{2n}{n}^{-1}.$$

By using the binomial formula in the integral below we compute

$$(8.46) \qquad \sum_{k=0}^{n} \binom{n}{k} (-1)^k \frac{x^k}{k+a} = \int_{0}^{\infty} (1 - xe^{-t})^n e^{-at} dt$$

for any $a > 0$. Then by integration

$$(8.47) \qquad \sum_{k=0}^{n} \binom{n}{k} (-1)^k x^k \ln(k+a)$$

$$= \int_{0}^{\infty} \left[(1-x)^n e^{-t} - (1-xe^{-t})^n e^{-at} \right] \frac{dt}{t}.$$

From here, for any $a, b > 0$

$$(8.48) \qquad \sum_{k=0}^{n} \binom{n}{k} (-1)^k x^k \ln \frac{k+a}{k+b}$$

$$= \int_{0}^{\infty} (1 - xe^{-t})^n (e^{-bt} - e^{-at}) \frac{dt}{t}$$

or

$$(8.49) \qquad \sum_{k=0}^{n} \binom{n}{k} (-1)^k x^k \ln \frac{k+a}{k+b}$$

$$= \int_{0}^{1} (z^{a-1} - z^{b-1})(1 - xz)^n \frac{dz}{\ln z}.$$

Note that the left hand side in (8.46) has an explicit representation given in (6.15). For convenience we repeat it here

(8.50)
$$\sum_{k=0}^{n}\binom{n}{k}\frac{(-1)^k x^k}{k+\lambda}$$

$$= (1-x)^n \sum_{j=0}^{n}\binom{n}{j}\frac{j!}{\lambda(\lambda+1)(\lambda+2)...(\lambda+j)}\left(\frac{x}{1-x}\right)^j$$

or, replacing x by $-x$

$$\sum_{k=0}^{n}\binom{n}{k}\frac{x^k}{k+\lambda}$$

$$= (1+x)^n \sum_{j=0}^{n}\binom{n}{j}\frac{(-1)^j j!}{\lambda(\lambda+1)(\lambda+2)...(\lambda+j)}\left(\frac{x}{1+x}\right)^j .$$

With $x = 1/2$ and any positive integer m we have from (8.50)

(8.51)
$$\sum_{k=0}^{n}\binom{n}{k}\frac{(-1)^k}{2^k(k+m)}$$

$$= \frac{1}{2^n}\sum_{j=0}^{n}\binom{n}{j}\frac{j!}{m(m+1)(m+2)...(m+j)}$$

$$= \frac{n!(m-1)!}{2^n(n+m)!}\sum_{k=0}^{n}\binom{n+m}{k}$$

(see Example 6.4).

For all $n \geq 0$, $a > 0$, $|x| \leq 1$, and $\mathrm{Re}(s) + n > 0$ the following representation extends (8.46). For the proof we expand $(1 - xe^{-t})^n$ inside the integral by the binomial formula and then integrate term by term. This gives

$$(8.52) \qquad \sum_{k=0}^{n} \binom{n}{k} \frac{x^k}{(k+a)^s}$$

$$= \frac{1}{\Gamma(s)} \int_0^\infty t^{s-1} e^{-at} (1 + xe^{-t})^n \, dt \ .$$

Changing s to $-s$ yields

$$(8.53) \qquad \sum_{k=0}^{n} \binom{n}{k} (-1)^k x^k (k+a)^s$$

$$= \frac{1}{\Gamma(-s)} \int_0^\infty e^{-at} \left(\frac{1 - xe^{-t}}{t} \right)^n \frac{dt}{t^{s-n+1}}$$

which is true for $\mathrm{Re}(s) < n$. This representation confirms that

$$(8.54) \qquad \sum_{k=0}^{n} \binom{n}{k} (-1)^k (k+a)^s = 0$$

for $s = 0, 1, \ldots, n-1$

$$(8.55) \qquad \sum_{k=0}^{n} \binom{n}{k} (-1)^{k-1} \frac{k}{(k+a)^3}$$

$$= \sum_{1 \leq i \leq j \leq n} \frac{1}{a^2 + (i+j)a + ij} \binom{n+a}{n}^{-1}$$

(see [23]).

The following identity is not a binomial transform, but is related to some of the above formulas and is interesting by itself

(8.56)
$$\sum_{k=1}^{n} \binom{n}{k} (-1)^k \left(\frac{1}{k+1} + \frac{1}{k+2} + \dots + \frac{1}{k+n} \right)$$
$$= \frac{1}{n} \left[1 - \binom{2n}{n}^{-1} \right].$$

Chapter 9

Identities Involving Harmonic Numbers

We start with an identity from Corollary 4.3. For any polynomial $f(x)$ with $\deg f \leq n$ we have

$$(9.1) \qquad \sum_{k=1}^{n} \binom{n}{k} (-1)^{k-1} \frac{f(x-k)}{k} = -f'(x) + f(x) H_n.$$

Setting here $f = 1$, we obtain the well-known

$$(9.2) \qquad \sum_{k=1}^{n} \binom{n}{k} \frac{(-1)^{k-1}}{k} = H_n.$$

By inversion

$$(9.3a) \qquad \sum_{k=0}^{n} \binom{n}{k} (-1)^{k-1} H_k = \frac{1}{n}$$

and more generally, in view of (1.10)

$$(9.3b) \qquad \sum_{k=0}^{n} \binom{n}{k} (-1)^{k-1} H_{k+p} = (-1)^p \Delta^p \frac{1}{n}.$$

Applying property (5.6) to (9.3a) we find

$$(9.4a) \qquad \sum_{k=1}^{n} \binom{n}{k} (-1)^{k-1} \frac{H_k}{k} = \sum_{k=1}^{n} \frac{1}{k^2} = H_n^{(2)}$$

and also by inversion

(9.4b)
$$\sum_{k=1}^{n} \binom{n}{k} (-1)^{k-1} H_n^{(2)} = \frac{H_n}{n}.$$

From (9.4a) again by (5.6)

(9.5)
$$\sum_{k=1}^{n} \binom{n}{k} (-1)^{k-1} \frac{H_k}{k^2} = \sum_{k=1}^{n} \frac{H_k^{(2)}}{k}.$$

From (9.2) and (5.6)

(9.6)
$$\sum_{k=1}^{n} \binom{n}{k} \frac{(-1)^{k-1}}{k^2} = \sum_{k=1}^{n} \frac{1}{k} H_k.$$

By inversion

(9.7)
$$\sum_{k=1}^{n} \binom{n}{k} (-1)^{k-1} \sum_{j=1}^{k} \frac{1}{j} H_j = \frac{1}{n^2}.$$

A simple computation shows that

(9.8)
$$\sum_{k=1}^{n} \frac{H_k}{k} = \frac{1}{2} \left(H_n^2 + H_n^{(2)} \right)$$

which can be used in the above identities. Thus (9.6) and (9.7) take the forms

(9.9)
$$\sum_{k=1}^{n} \binom{n}{k} \frac{(-1)^{k-1}}{k^2} = \frac{1}{2} \left(H_n^2 + H_n^{(2)} \right)$$

(9.10)
$$\sum_{k=1}^{n} \binom{n}{k} (-1)^{k-1} \left(H_n^2 + H_n^{(2)} \right) = \frac{2}{n^2}$$

(see also Example 3.8).

Further, from (9.9) and (5.6)

(9.11a)
$$\sum_{k=1}^{n} \binom{n}{k} \frac{(-1)^{k-1}}{k^3} = \frac{1}{2} \sum_{k=1}^{n} \frac{1}{k} \left(H_n^2 + H_n^{(2)} \right)$$

which can be written also in the form

(9.11b)
$$\sum_{k=1}^{n} \binom{n}{k} \frac{(-1)^{k-1}}{k^3} = \frac{1}{6} \left(H_n^3 + 3 H_n H_n^{(2)} + 2 H_n^{(3)} \right)$$

and also in the form

(9.11c)
$$\sum_{k=1}^{n} \binom{n}{k} \frac{(-1)^{k-1}}{k^3} = \sum \frac{1}{k_1 k_2 k_3}$$

where the sum is for $1 \le k_1 \le k_2 \le k_3 \le n$. This extends to

(9.12)
$$\sum_{k=1}^{n} \binom{n}{k} \frac{(-1)^{k-1}}{k^m} = \sum \frac{1}{k_1 k_2 \ldots k_m}$$

with summation for $1 \le k_1 \le k_2 \le \ldots \le k_m \le n$ (see Example 5.4).

(9.13)
$$\sum_{k=1}^{n} \binom{n}{k} \frac{(-1)^{k-1}}{k^4}$$

$$= \frac{1}{24} \left(H_n^4 + 6 H_n^2 H_n^{(2)} + 3 \left(H_n^{(2)} \right)^2 + 8 H_n H_n^{(3)} + 6 H_n^{(4)} \right).$$

Writing $H_j = H_{j-1} + \dfrac{1}{j}$ in (9.7) and using (9.4b)

(9.14)
$$\sum_{k=1}^{n} \binom{n}{k} (-1)^{k-1} \sum_{j=1}^{k} \frac{1}{j} H_{j-1} = \frac{1}{n^2} - \frac{H_n}{n}.$$

From (9.10) and (5.6)

(9.15) $$\sum_{k=1}^{n}\binom{n}{k}\frac{(-1)^{k-1}}{k}\left(H_n^2+H_n^{(2)}\right)=2\sum_{k=1}^{n}\frac{1}{k^3}=2H_n^{(3)}$$

(cf. Larsen [46], p. 190). By inversion also

(9.16) $$\sum_{k=1}^{n}\binom{n}{k}(-1)^{k-1}H_k^{(3)}=\frac{1}{2n}\left(H_n^2+H_n^{(2)}\right).$$

For any x

(9.17) $$\sum_{k=1}^{n}\binom{n}{k}(-1)^{k-1}\sum_{m=1}^{k}\frac{x^m}{m}=\frac{1-(1-x)^n}{n}$$

(Riordan [59], p. 27; Schwatt [61], p. 54). This identity is easily verified by differentiation. By inversion also

(9.18) $$\sum_{k=1}^{n}\binom{n}{k}(-1)^{k-1}\frac{1-(1-x)^k}{k}=\sum_{m=1}^{n}\frac{x^m}{m}.$$

With $x=-1$ in (9.17)

(9.19) $$\sum_{k=1}^{n}\binom{n}{k}(-1)^{k-1}\sum_{m=1}^{k}\frac{(-1)^m}{m}=\frac{1-2^n}{n}$$

or using the notation H_k^- (see (2.14))

(9.20) $$\sum_{k=1}^{n}\binom{n}{k}(-1)^{k-1}H_k^-=\frac{2^n-1}{n}.$$

Inversion yields

(9.21) $$\sum_{k=1}^{n}\binom{n}{k}(-1)^{k-1}\frac{2^k-1}{k}=H_n^-.$$

From (9.20) by (5.6) and (5.7) correspondingly,

$$(9.22) \qquad \sum_{k=1}^{n}\binom{n}{k}(-1)^{k-1}\frac{H_k^-}{k}=\sum_{k=1}^{n}\frac{2^k-1}{k^2}=\sum_{k=1}^{n}\frac{2^k}{k^2}-H_n^{(2)}$$

$$(9.23) \qquad \sum_{k=1}^{n}\binom{n}{k}(-1)^{k-1}\frac{H_k^-}{k+1}=\frac{1}{n+1}\left\{\sum_{k=1}^{n}\frac{2^k}{k}-H_n\right\}.$$

Starting from the obvious

$$(9.24) \qquad \sum_{k=1}^{n}\binom{n}{k}(-1)^k x^k = (1-x)^k - 1$$

by integration we find

$$(9.25) \qquad \sum_{k=0}^{n}\binom{n}{k}(-1)^k\frac{x^{k+1}}{k+1}=\frac{1-(1-x)^{n+1}}{n+1}.$$

Applying (5.6) to (9.24)

$$(9.26) \qquad \sum_{k=1}^{n}\binom{n}{k}(-1)^k\frac{x^k}{k}=\sum_{k=1}^{n}\frac{(1-x)^k}{k}-H_n$$

and changing x to $-x$,

$$(9.27) \qquad \sum_{k=1}^{n}\binom{n}{k}\frac{x^k}{k}=\sum_{k=1}^{n}\frac{(1+x)^k}{k}-H_n.$$

With $x=1$ and $x=-2$ in (9.27)

$$(9.28) \qquad \sum_{k=1}^{n}\binom{n}{k}\frac{1}{k}=\sum_{k=1}^{n}\frac{2^k}{k}-H_n$$

$$(9.29) \quad \sum_{k=1}^{n} \binom{n}{k} \frac{(-2)^k}{k} = \sum_{k=1}^{n} \frac{(-1)^k}{k} - H_n = \sum_{k=1}^{n} \frac{(-1)^k - 1}{k}.$$

From (9.27) by using (5.6)

$$(9.30) \quad \sum_{k=1}^{n} \binom{n}{k} \frac{x^k}{k^2} = \sum_{k=1}^{n} \frac{1}{k} \sum_{j=1}^{k} \frac{(1+x)^j}{j} - \sum_{k=1}^{n} \frac{H_k}{k}.$$

With $x = -1$ this becomes (9.6). With $x = 1$ we have

$$(9.31) \quad \sum_{k=1}^{n} \binom{n}{k} \frac{1}{k^2} = \sum_{k=1}^{n} \frac{1}{k} \sum_{j=1}^{k} \frac{2^j}{j} - \sum_{k=1}^{n} \frac{H_k}{k}.$$

By applying (5.7) to (9.3) we find

$$(9.32) \quad \sum_{k=1}^{n} \binom{n}{k} (-1)^{k-1} \frac{H_k}{k+1} = \frac{H_n}{n+1}$$

which shows that the sequence $\dfrac{H_k}{k+1}$ has the invariance property (3.57).

Its generating function

$$(9.33) \quad f(t) = \frac{1}{2t} \ln^2(1-t) = \sum_{k=1}^{\infty} \frac{H_k}{k+1} t^k$$

satisfies the functional equation

$$\frac{1}{1-t} f\left(\frac{-t}{1-t}\right) = -f(t)$$

(see Example 3.3 and Section 3.5).

Writing $H_{k+1} = H_k + \dfrac{1}{k+1}$ we find also

$$(9.34) \qquad \sum_{k=0}^{n} \binom{n}{k} (-1)^{k-1} H_{k+1} = \frac{1}{n} - \frac{1}{n+1}$$

(this follows independently from property (1.10)). Applying (5.7) gives

$$(9.35) \qquad \sum_{k=0}^{n} \binom{n}{k} (-1)^{k} \frac{H_{k+1}}{k+1} = \frac{1}{(n+1)^2} \,.$$

Inverting this we get

$$(9.36) \qquad \sum_{k=0}^{n} \binom{n}{k} \frac{(-1)^{k}}{(k+1)^2} = \frac{H_{n+1}}{n+1} \,.$$

Next, from [21]

$$(9.37) \qquad \sum_{k=0}^{n} \binom{n}{k} (-1)^{k-1} H_{k+p} = \frac{1}{n} \binom{n+p}{p}^{-1}$$

$$= \frac{p!}{n(n+1)...(n+p)}$$

(cf. (9.3b)). This can be written also in the form

$$(9.38) \qquad \sum_{k=0}^{n} \binom{n}{k} (-1)^{k-1} \left(H_{k+p} - H_p \right) = \frac{p!}{n(n+1)...(n+p)}$$

where the summation starts, in fact, from $k=1$. By inversion

$$(9.39) \qquad \sum_{k=1}^{n} \binom{n}{k} \frac{(-1)^{k-1}}{k(k+1)...(k+p)} = \frac{1}{p!} \left(H_{n+p} - H_p \right).$$

Further,

$$(9.40) \qquad \sum_{k=0}^{n} \binom{n}{k}(-1)^k (H_{k+p}^2 + H_{k+p}^{(2)}) = \frac{2}{n}(H_{n-1} - H_{n+p})\binom{n+p}{p}^{-1}$$

$$(9.41) \qquad \sum_{k=0}^{n} \binom{n}{k}(-1)^k (H_{k+p}^2 - H_{k+p}^{(2)}) = \frac{2}{n}(H_{n-1} - H_{p})\binom{n+p}{p}^{-1}.$$

When $p = 0$ (9.40) becomes (9.10) and (9.41) becomes

$$(9.42) \qquad \sum_{k=0}^{n} \binom{n}{k}(-1)^k (H_k^2 - H_k^{(2)}) = \frac{2}{n}H_{n-1}$$

which together with (9.10) provides

$$(9.43) \qquad \sum_{k=0}^{n} \binom{n}{k}(-1)^k H_k^2 = \frac{1}{n}H_{n-1} - \frac{1}{n^2}.$$

We have also

$$(9.44) \qquad \sum_{k=0}^{n} \binom{n}{k}(-1)^k H_k^3 = \frac{1}{2n}\left(5H_{n-1}^{(2)} + \frac{4}{n}H_{n-1} - H_{n-1}^2 - \frac{2}{n^2}\right)$$

(from [74]).

$$(9.45) \qquad \sum_{k=0}^{n} \binom{n}{k}(-1)^k H_k H_k^{(2)} = \frac{1}{2n}\left(H_{n-1}^2 - H_{n-1}^{(2)} - \frac{2}{n^2}\right).$$

$$(9.46) \qquad \sum_{k=0}^{n} \binom{n}{k}(-1)^k \frac{H_{k+p}}{k+p} = \left(H_{n+p} - H_n\right)\frac{1}{p}\binom{n+p}{n}^{-1}.$$

Extending (9.2)

$$(9.47) \qquad \sum_{k=0}^{n} \binom{n}{k}(-1)^k \left\{\sum_{m=1}^{k}\frac{1}{k+m}\right\} = \frac{1}{n}\left(1 - \frac{n!m!}{(n+m)!}\right).$$

(9.48) $$\sum_{k=1}^{n}\binom{n}{k}(-1)^{k-1}H_{2k} = \frac{1}{2n} + \frac{2.4....(2n-2)}{3.5....(2n-1)}$$

(9.49) $$\sum_{k=1}^{n}\binom{n}{k}(-1)^{k-1}(2H_{2k}-H_{k}) = \binom{2n}{n}^{-1}\frac{2^{2n}}{n}.$$

With the notation

(9.50) $$h_{k} = H_{2k} - \frac{1}{2}H_{k} = 1 + \frac{1}{3} + ... + \frac{1}{2k-1}$$

we can write (9.49) in the form

(9.51) $$\sum_{k=1}^{n}\binom{n}{k}(-1)^{k-1}h_{k} = \binom{2n}{n}^{-1}\frac{2^{2n-1}}{n}$$

and by inversion

(9.52) $$\sum_{k=1}^{n}\binom{n}{k}(-1)^{k-1}\binom{2k}{k}^{-1}\frac{2^{2k-1}}{k} = h_{n}$$

or

(9.53) $$\sum_{k=1}^{n}\binom{n}{k}(-1)^{k-1}\frac{2^{2k}}{k}\binom{2k}{k}^{-1} = 2H_{2n} - H_{n}.$$

For every positive integer n and every two complex numbers x, y we have (see [4] and [44])

(9.54) $$\sum_{k=1}^{n}\binom{n}{k}H_{k}x^{n-k}y^{k}$$

$$= (x+y)^n H_n - \left[x(x+y)^{n-1} + \frac{x^2}{2}(x+y)^{n-2} + \ldots + \frac{x^n}{n} \right].$$

In particular, with $x=1$

(9.55)
$$\sum_{k=1}^{n} \binom{n}{k} H_k y^k$$

$$= (y+1)^n H_n - \left[(y+1)^{n-1} + \frac{1}{2}(y+1)^{n-2} + \ldots + \frac{1}{n} \right]$$

and with $x=y=1$

(9.56)
$$\sum_{k=1}^{n} \binom{n}{k} H_k = 2^n \left(H_n - \sum_{k=1}^{n} \frac{1}{k2^k} \right).$$

Also from (3.33) (cf. [4])

(9.57)
$$\sum_{k=1}^{n} \binom{n}{k} H_k k^m x^k = a(m,n,x) H_n - \sum_{k=1}^{n-1} \frac{a(m,k,x)}{n-k}$$

where $n>1$ and

(9.58)
$$a(m,n,x) = \left(x \frac{d}{dx} \right)^m (x+1)^n = \sum_{k=0}^{n} \binom{n}{k} k^m x^k$$

also in the form

(9.59)
$$a(m,n,x) = \sum_{j=0}^{n} \binom{n}{j} S(m,j) j! x^j (1+x)^{n-j}.$$

This is computed by using the rule

$$(9.60) \qquad \left(x\frac{d}{dx}\right)^m f(x) = \sum_{j=0}^n \binom{n}{j} S(m,j)\, x^j f^{(j)}(x).$$

When $x = -1$, $a(m,n,-1) = (-1)^n n! S(m,n)$, and (9.57) becomes

$$(9.61) \qquad \sum_{k=0}^n \binom{n}{k} (-1)^k k^m H_k$$

$$= (-1)^n H_n S(m,n) n! - \sum_{k=0}^{n-1} \frac{(-1)^k}{n-k} S(m,k) k!\,.$$

For any $p \geq 1$

$$(9.62) \qquad \sum_{k=1}^n \binom{n}{k}\binom{p}{k} H_k = \binom{n+p}{n}(H_p + H_n - H_{p+n})$$

(see [8]).

For all natural numbers $m, p \geq n$

$$(9.63) \qquad \sum_{k=1}^n \binom{n}{k}(-1)^k \binom{p-k}{m}(H_{p-k} - H_m)$$

$$= \binom{p-n}{m-n}(H_{p-n} - H_{m-n})$$

and more generally (from [63])

$$(9.64) \qquad \sum_{k=1}^n \binom{n}{k}(-1)^k \binom{p-k}{m} P(r, p-k, m)$$

$$= \binom{p-n}{m-n} P(r, p-n, m-n)$$

where $P(r,k,m) = P_r(H_k^{(1)} - H_m^{(1)}, ..., H_k^{(r)} - H_m^{(r)})$, and P_r is defined in terms of the Bell polynomials Y_r (Comtet [22]), namely,

$$P_r(x_1, ..., x_r) = (-1)^r Y_r(-0!x_1, -1!x_2, ..., -(r-1)!x_r).$$

The following six entries are from [20], where other similar formulas and more details can be found. For any $m \geq 1$ define

(9.65) $$H_0(x) = 0, \quad H_m(x) = \sum_{k=1}^{m} \frac{1}{k+x}.$$

Then for $p \geq 2$

(9.66) $$\sum_{k=1}^{n} \binom{n}{k} (-1)^{k-1} \binom{x+k}{k} \binom{p+k}{k}^{-1} H_k(x)$$

$$= \frac{(p-x)_n}{(p+1)_n} \left(H_{n+p-1}(-x) - H_{p-1}(-x) \right).$$

In particular, for $x = 0$

(9.67) $$\sum_{k=1}^{n} \binom{n}{k} (-1)^{k-1} \binom{p+k}{k}^{-1} H_k$$

$$= \frac{p}{p+n} \left(H_{n+p-1} - H_{p-1} \right)$$

(9.68) $$\sum_{k=1}^{n} \binom{n}{k} (-1)^{k} \binom{p+k}{k}^{-1} k H_k$$

$$= \frac{pn}{(p+n)(p+n-1)} \left(H_{p+n-2} - H_{p-1} - 1 \right)$$

(9.69) $$\sum_{k=1}^{n}\binom{n}{k}(-1)^{k}\binom{p+k}{k}^{-1}\left(H_{p+k}-H_{k}\right)=\frac{p}{p+n}H_{p+n}$$

(9.70) $$\sum_{k=1}^{n}\binom{n}{k}(-1)^{k}\binom{p+k}{k}^{-1}k\left(H_{p+k}-H_{k}\right)$$

$$=\frac{pn}{(p+n)(p+n-1)}\left(1-H_{p+n}\right).$$

For $p\geq 3$

(9.71) $$\sum_{k=1}^{n}\binom{n}{k}(-1)^{k}\binom{p+k}{k}^{-1}k^{2}\left(H_{p+k}-H_{k}\right)$$

$$=\frac{pn(1+p-2n)+pn(n-p)H_{p+n}}{(p+n)(p+n-1)(p+n-2)}.$$

For completeness we include here results from Corollary 6.8. Let $\{c_{k}\}$ and $\{d_{k}\}$ be any two sequences related as in (6.3). Then

(9.72) $$\sum_{k=0}^{n}\binom{n}{k}(-1)^{k-1}H_{k}c_{k}=(-1)^{n-1}H_{n}d_{n}+\sum_{m=0}^{n-1}\frac{(-1)^{m}d_{m}}{n-m}.$$

In particular,

(9.73) $$\sum_{k=0}^{n}\binom{n}{k}(-1)^{k-1}\frac{H_{k}^{2}}{k}=H_{n}H_{n}^{(2)}-\sum_{m=0}^{n-1}\frac{H_{m}^{(2)}}{n-m}$$

(9.74) $$\sum_{k=0}^{n}\binom{n}{k}(-1)^{k-1}\frac{H_{k}^{2}}{k+1}=\frac{H_{n}^{2}}{n+1}-\sum_{m=0}^{n-1}\frac{H_{m}}{(n-m)(m+1)}$$

(9.75) $$\sum_{k=0}^{n}\binom{n}{k}(-1)^{k-1}H_{k}F_{k}=H_{n}F_{n}-\sum_{m=0}^{n-1}\frac{F_{m}}{n-m}$$

(9.76) $\quad \sum_{k=0}^{n} \binom{n}{k} (-1)^{k-1} H_k F_{2k} = (-1)^{n-1} H_n F_n + \sum_{m=0}^{n-1} \frac{(-1)^m F_m}{n-m}$

where F_n are the Fibonacci numbers. With the Bernoulli numbers B_n and with the Laguerre polynomials $L_n(x)$ we have likewise

(9.77) $\quad \sum_{k=0}^{n} \binom{n}{k} H_k B_n = (-1)^n H_n B_n - \sum_{m=0}^{n-1} \frac{(-1)^m B_m}{n-m}$

(9.78) $\quad \sum_{k=0}^{n} \binom{n}{k} (-1)^k H_k L_k(x) = \frac{x^n}{n!} H_n - \sum_{m=0}^{n-1} \frac{x^m}{m!(n-m)} \quad.$

Chapter 10

Transforms of Binomial Coefficients

Many entries in this chapter are from the books [31] and [57].
For $0 \le x \le y$

$$(10.1) \qquad \sum_{k=0}^{n} \binom{n}{k} (-1)^k \binom{x}{k} \binom{y}{k}^{-1} = \binom{y-x}{n} \binom{y}{n}^{-1}$$

$$(10.2) \qquad \sum_{k=0}^{n} \binom{n}{k} \binom{x}{k} \binom{y+k}{k}^{-1} = \binom{n+y+x}{n} \binom{n+y}{n}^{-1}$$

$$(10.3) \qquad \sum_{k=0}^{n} \binom{n}{k} (-1)^k \binom{k+y}{x}^{-1} = \frac{x}{n+x} \binom{n+y}{y-x}^{-1}$$

with inverse

$$(10.4) \qquad \sum_{k=0}^{n} \binom{n}{k} (-1)^k \frac{x}{k+x} \binom{k+y}{y-x}^{-1} = \binom{n+y}{x}^{-1}.$$

When $y = x$ equation (10.4) becomes (8.28b) and (10.3) becomes (8.29).

$$(10.5) \qquad \sum_{k=0}^{n} \binom{n}{k} (-1)^k \frac{(a)_k}{(b)_k} = \prod_{j=0}^{n-1} \left(1 - \frac{a}{b-j} \right)$$

where $(x)_n = x(x-1)...(x-n+1)$, $(x)_0 = 1$ is the falling factorial.

139

$$(10.6) \qquad \sum_{k=0}^{n} \binom{n}{k} \binom{x+k}{k} \binom{y+k}{k}^{-1} = \frac{(y-x)^{(n)}}{(y+1)^{(n)}}$$

where $x^{(n)} = x(x+1)...(x+n-1)$, $x^{(0)} = 1$ is the rising factorial.

For any p, y

$$(10.7) \qquad \sum_{k=0}^{n} \binom{n}{k} (-1)^k \binom{y-k}{p} = \binom{y-n}{y-p}$$

$$(10.8) \qquad \sum_{k=0}^{n} \binom{n}{k} (-1)^k \binom{y+k}{p} = (-1)^n \binom{y}{p-n}.$$

In particular, when $y = p$

$$(10.9) \qquad \sum_{k=0}^{n} \binom{n}{k} (-1)^k \binom{p+k}{p} = \sum_{k=0}^{n} \binom{n}{k} (-1)^k \binom{p+k}{k}$$

$$= (-1)^n \binom{p}{p-n} = (-1)^n \binom{p}{n}.$$

The next five identities are known as Vandermonde convolutions, see [31] and [37].

$$(10.10a) \qquad \sum_{k=0}^{n} \binom{n}{k} \binom{p}{k+h} = \binom{n+p}{n+h}$$

with inversion

$$(10.10b) \qquad \sum_{k=0}^{n} \binom{n}{k} \binom{k+p}{k+h} (-1)^k = (-1)^n \binom{p}{n+h}.$$

Changing h to $-h$ in (10.10a) gives also

(10.10c)
$$\sum_{k=0}^{n}\binom{n}{k}\binom{p}{k-h}=\binom{n+p}{n-h}.$$

In particular, when $h=0$

(10.10d)
$$\sum_{k=0}^{n}\binom{n}{k}\binom{p}{k}=\binom{n+p}{n}$$

(10.11)
$$\sum_{k=0}^{n}\binom{n}{k}\binom{a+b}{a-k}=\binom{a+b+n}{a}=\binom{a+b+n}{b+n}.$$

In terms of the falling factorial $(k)_m = k(k-1)...(k-m+1)$ we have (cf. (6.10))

(10.12)
$$\sum_{k=0}^{n}\binom{n}{k}(k)_m = 2^{n-m}(n)_m.$$

Further,

(10.13)
$$\sum_{k=0}^{n}\binom{n}{k}(-1)^k\binom{2k}{p}=(-1)^n\binom{n}{p-n}2^{2n-p}$$

(10.14)
$$\sum_{k=0}^{n}\binom{n}{k}\binom{k}{p-k}2^{2k}=2^p\binom{2n}{p}$$

(10.15)
$$\sum_{k=0}^{n}\binom{n}{k}\binom{\alpha}{k}x^{n-k}y^k = \sum_{k=0}^{n}\binom{n}{k}\binom{\alpha+k}{k}(x-y)^{n-k}y^k.$$

This is Ljunggren's identity (see [7] for comments). It is listed as entry 3.18 in [31]. The identity can be written as a binomial transform by setting $z = y/x$

(10.16) $$\sum_{k=0}^{n}\binom{n}{k}\binom{\alpha}{k}z^k = (1-z)^n \sum_{k=0}^{n}\binom{n}{k}\binom{\alpha+k}{k}\left(\frac{z}{1-z}\right)^k.$$

Renaming the variables in Ljunggren's identity $x - y = t$ and $y = s$ we can write it in the form

(10.17) $$\sum_{k=0}^{n}\binom{n}{k}\binom{\alpha}{k}(t+s)^{n-k}s^k = \sum_{k=0}^{n}\binom{n}{k}\binom{\alpha+k}{k}s^k t^{n-k}.$$

Also from [7]

(10.18) $$\sum_{k=0}^{n}\binom{n}{k}\binom{\alpha+k}{k}x^k = \sum_{k=0}^{n}\binom{\alpha}{n-k}\binom{\alpha+k}{k}(-1)^{n-k}(x+1)^k.$$

When $\alpha = n$ this turns into the "curious identity" of Simons ([7], [52], [60]).

$$\sum_{k=0}^{n}\binom{n}{k}\binom{n+k}{k}x^k = \sum_{k=0}^{n}\binom{n}{n-k}\binom{n+k}{k}(-1)^{n-k}(x+1)^k.$$

The polynomials on the left hand side are known as the Delannoy polynomials $D_n(x)$. The numbers $D_n(1)$ are known as the central Delannoy numbers. Similar to this is the representation

$$\sum_{k=0}^{n}\binom{n}{k}\binom{n+k}{k}\frac{(x-1)^k}{2^k} = P_n(x)$$

where $P_n(x), n = 0, 1, \dots$ are the Legendre polynomials.

The last two identities are not binomial transforms and are listed here for convenience.

When $x = -1$ in (10.18) it turns into (10.9) with $\alpha = p$.

The next identity is entry (3.17) from [31]

(10.19) $$\sum_{k=0}^{n}\binom{n}{k}\binom{\alpha}{k}x^k = \sum_{k=0}^{n}\binom{n}{k}\binom{n+\alpha-k}{n}(x-1)^k.$$

For $x=2$ this becomes

(10.20) $$\sum_{k=0}^{n}\binom{n}{k}\binom{\alpha}{k}2^k = \sum_{k=0}^{n}\binom{n}{k}\binom{n+\alpha-k}{n}$$

$$= \sum_{j=0}^{n}\binom{n}{j}\binom{\alpha+j}{n}$$

When $\alpha = m$ is an integer the numbers defined here are called the (general) Delannoy numbers

(10.21) $$D(m,n) = \sum_{j=0}^{n}\binom{n}{j}\binom{m}{j}2^j$$

$$= \sum_{k=0}^{n}\binom{n}{k}\binom{n+m-k}{n} = \sum_{j=0}^{n}\binom{n}{j}\binom{m+j}{n}$$

(see [22], p. 81, for their combinatorial interpretation).

Differentiating with respect to x in (10.19) and then setting $x=1$ we find

(10.22) $$\sum_{k=0}^{n}\binom{n}{k}\binom{\alpha}{k}k = n\binom{n+\alpha-1}{n} = \alpha\binom{n+\alpha-1}{n-1}$$

(10.23) $$\sum_{k=0}^{n}\binom{n}{k}\binom{k}{q}x^k = \binom{n}{q}x^q(1+x)^{n-q}.$$

With $x=1$ here we have

(10.24) $$\sum_{k=0}^{n}\binom{n}{k}\binom{k}{q} = 2^{n-q}\binom{n}{q}.$$

Changing x to $-x$ in (10.23) yields

$$(10.25) \qquad \sum_{k=0}^{n} \binom{n}{k} (-1)^k \binom{k}{q} x^k = (-1)^q \binom{n}{q} x^q (1-x)^{n-q} .$$

With $x = 2$ this becomes

$$(10.26a) \qquad \sum_{k=0}^{n} \binom{n}{k} (-1)^k \binom{k}{q} 2^k = (-1)^n 2^q \binom{n}{q}$$

and with $x = \dfrac{1}{2}$ we have

$$(10.26b) \qquad \sum_{k=0}^{n} \binom{n}{k} (-1)^k \binom{k}{q} \frac{1}{2^k} = \frac{(-1)^q}{2^n} \binom{n}{q} .$$

Integration in (10.25) with respect to x from 0 to 1 yields

$$(10.27) \qquad \sum_{k=0}^{n} \binom{n}{k} \binom{k}{q} \frac{(-1)^k}{k+1} = (-1)^q \binom{n}{q} \mathrm{B}(q+1, n-q+1)$$

where $\mathrm{B}(x, y)$ is Euler's Beta function. Using its property

$$\mathrm{B}(q+1, n-q+1) = \frac{q!(n-q)!}{(n+1)!}$$

we can write

$$(10.28) \qquad \sum_{k=0}^{n} \binom{n}{k} \binom{k}{q} \frac{(-1)^k}{k+1} = \frac{(-1)^q}{n+1}$$

$$(10.29) \qquad \sum_{k=0}^{n} \binom{n}{k} \binom{k}{q} \frac{(-x)^k}{k!} = \frac{(-x)^q}{q!} L_{n-q}^q(x)$$

where $L_n^\alpha(x)$ are the generalized Laguerre polynomials (see (3.50)).

Next we have

$$(10.30) \qquad \sum_{k=0}^{n} \binom{n}{k} (-1)^k \binom{2k}{k} \frac{1}{2^k} = \frac{1}{4^p} \binom{2p}{p}$$

for $n = 2p$ (n even) and

$$(10.31) \qquad \sum_{k=0}^{n} \binom{n}{k} (-1)^k \binom{2k}{k} \frac{1}{2^k} = 0$$

for n odd. Also

$$(10.32) \qquad \sum_{k=0}^{n} \binom{n}{k} (-1)^k \binom{2k}{k} \frac{1}{4^k(1-2k)} = \frac{2n+1}{4^n} \binom{2n}{n}$$

$$(10.33) \qquad \sum_{k=0}^{n} \binom{n}{k} (-1)^k \binom{2k}{k} \frac{2k+1}{4^k} = \frac{1}{4^n(1-2n)} \binom{2n}{n}$$

$$(10.34) \qquad \sum_{k=0}^{n} \binom{n}{k} (-1)^k \binom{2k}{k} \frac{z^k}{4^k} = \frac{1}{4^n} \sum_{i=0}^{n} \binom{2n-2i}{n-i}\binom{2i}{i}(1-z)^i.$$

With $z = 1$, $z = 4$, and $z = -4$ we obtain correspondingly the next three identities

$$(10.35a) \qquad \sum_{k=0}^{n} \binom{n}{k} (-1)^k \binom{2k}{k} \frac{1}{4^k} = \frac{1}{4^n} \binom{2n}{n}$$

$$(10.35b) \qquad \sum_{k=0}^{n} \binom{n}{k} (-1)^k \binom{2k}{k} = \frac{1}{4^n} \sum_{i=0}^{n} \binom{2n-2i}{n-i}\binom{2i}{i}(-3)^i$$

$$(10.35c) \qquad \sum_{k=0}^{n} \binom{n}{k} \binom{2k}{k} = \frac{1}{4^n} \sum_{i=0}^{n} \binom{2n-2i}{n-i}\binom{2i}{i} 5^i$$

(10.36)
$$\sum_{k=0}^{n}\binom{n}{k}(-1)^k\binom{2k}{k}\binom{k+p}{k}^{-1}\frac{1}{4^k}$$

$$=\frac{1}{4^n}\binom{2n+2p}{n+p}\binom{2p}{p}^{-1}$$

(10.37) $$\sum_{k=0}^{n}\binom{n}{k}(-1)^k\binom{kx}{p}=(-1)^n\frac{n!}{p!}\sum_{k=n}^{p}s(p,k)S(k,n)x^k$$

(see Example 7.1 in Chapter 7).

(10.38)
$$\sum_{k=1}^{n}\binom{n}{k}(-1)^{k-1}\binom{2k}{k}^{-1}\frac{2^{2k}}{k}=2H_{2n}-H_n$$

(the inverse of (9.49))

(10.39)
$$\sum_{k=0}^{n}\binom{n}{k}(-1)^k\binom{2k}{k}^{-1}\frac{2^{2k}}{2k+1}=\frac{1}{2n+1}$$

which is the inverse of (8.43) (see (1.13)). Also

(10.40)
$$\sum_{k=0}^{n}\binom{n}{k}(-1)^k\binom{2k}{k}^{-1}4^k=\frac{1}{1-2n}$$

which is the inverse of (8.44).

An important convolution property follows from equation (10.23) by setting there $x=-1$

(10.41a)
$$\sum_{k=0}^{n}\binom{n}{k}(-1)^k\binom{k}{q}=(-1)^q\delta_{nq}$$

where δ_{nq} is the Kronecker symbol. This identity follows also by inversion from the obvious identity

$$(10.41\text{b}) \qquad \sum_{k=0}^{n} \binom{n}{k} (-1)^k (-1)^q \delta_{kq} = \binom{n}{q} .$$

$$(10.42) \qquad \sum_{k=0}^{n} \binom{n}{k}\binom{x}{k} \frac{1}{k+1} = \frac{1}{n+1} \binom{n+x+1}{n}$$

$$(10.43) \qquad \sum_{k=0}^{n} \binom{n}{k}\binom{x}{k+y}^{-1} = \frac{x+1}{x-n+1} \binom{x-n}{y}$$

$$(10.44) \qquad \sum_{k=0}^{n} \binom{n}{k} (-1)^k \binom{k+p}{k}^{-1} k^r$$

$$= \binom{-p-1}{n}^{-1} \sum_{k=0}^{r} \binom{-p}{n-k} S(r,k) k! .$$

Differentiating (10.9) with respect to p we obtain

$$(10.45) \qquad \sum_{k=0}^{n} \binom{n}{k} (-1)^k \binom{p+k}{p} \{\psi(p+k+1)-\psi(p+1)\}$$

$$= (-1)^n \binom{p}{n} \{\psi(p+1)-\psi(p-n+1)\}$$

where ψ is the digamma function and the right hand side is well-defined for $p \geq n$.

For integers $0 \leq p < n$ we have

$$(10.46) \qquad \sum_{k=0}^{n} \binom{n}{k} (-1)^k \binom{p+k}{p} \{\psi(p+k+1)-\psi(p+1)\}$$

$$= \frac{(-1)^{p+1}}{n-p} \binom{n}{p}^{-1} .$$

When $p \geq 0$ is an integer we find from here the binomial transform of the hyperharmonic numbers

$$(10.47) \qquad H_{n,p+1} = \binom{p+k}{p}\{\psi(p+k+1) - \psi(p+1)\}$$

$$= \binom{p+k}{p}\{H_{p+k} - H_p\}.$$

Namely, for $0 \leq p < n$ we have

$$(10.48) \qquad \sum_{k=0}^{n}\binom{n}{k}(-1)^k H_{n,p+1} = \frac{(-1)^{p+1}}{n-p}\binom{n}{p}^{-1}.$$

Further,

$$(10.49) \qquad \sum_{k=0}^{n}\binom{n}{k}\binom{x}{k}^{-1}k = \frac{n(x+1)}{(n-x-1)(n-x-2)}$$

$$(10.50) \qquad \sum_{k=0}^{n}\binom{n}{k}\binom{x}{k}^{-1}k^2 = \frac{n(x+1)(n+x+1)}{(n-x-1)(n-x-2)(n-x-3)}$$

$$(10.51) \qquad \sum_{k=0}^{n}\binom{n}{k}(-1)^k\binom{2k}{k}^{-1}\binom{x+k}{k}\frac{4^k x}{x+k} = (-1)^n\binom{x}{n}^{-1}\binom{2x}{2n}$$

$$(10.52) \qquad \sum_{k=0}^{n}\binom{n}{k}(-1)^k\binom{2k}{k}^{-1}\binom{x+k}{k}\frac{4^k}{2k+1}$$

$$= \frac{4^n}{2n+1}\binom{n-x-1/2}{n}\binom{2n}{n}^{-1}.$$

(10.53)
$$\sum_{k=0}^{n} \binom{n}{k}\binom{k+m}{k}^{-1} x^k$$

$$= \left[m + (x+1)^{n+m} - \sum_{j=0}^{m} \binom{n+m}{j} x^j \right] \binom{n+m}{n}^{-1} x^{-m}$$

(10.54)
$$\sum_{k=0}^{n} \binom{n}{k}(-1)^k \binom{k}{p}\binom{k}{m} = (-1)^n \binom{n}{m}\binom{m}{n-p}$$

(10.55)
$$\sum_{k=0}^{n} \binom{n}{k}(-1)^k \binom{m-a-1}{k+p}\binom{a}{m-p-k}$$

$$= (-1)^p \binom{m+n}{m-1}\binom{a}{m}$$

and for $n \geq 1$

(10.56)
$$\sum_{k=0}^{n} \binom{n}{k}(-1)^k \binom{k+x}{x}\binom{k+y}{y}^{-1} = \prod_{k=0}^{n-1}\left(1 - \frac{x+1}{y+1-k} \right).$$

Chapter 11

Transforms of Special Numbers and Polynomials

First we consider power sums. For a positive integer $p > 0$ define

$$(11.1) \qquad S_k(p) = 1^k + 2^k + \ldots + p^k$$

$$(11.2) \qquad S_k^-(p) = 1^k - 2^k + \ldots + (-1)^{p-1} p^k.$$

Then we have (Nielsen [53], Riordan [59], p. 159, Turner [72])

$$(11.3) \qquad \sum_{k=0}^{n} \binom{n}{k} S_k(p) = S_n(p) + (p+1)^n - 1$$

$$(11.4) \qquad \sum_{k=0}^{n} \binom{n}{k} (-1)^k S_k(p) = (-1)^n \left(S_n(p) - p^n \right)$$

$$(11.5) \qquad \sum_{k=0}^{n} \binom{n}{k} S_k^-(p) = -S_n^-(p) + (-1)^{n-1}(p+1)^n + 1$$

$$(11.6) \qquad \sum_{k=0}^{n} \binom{n}{k} (-1)^k S_k^-(p) = (-1)^{n-1} S_n^-(p) - p^n.$$

For any nonnegative integer q we define the sequence of sums

$$(11.7) \qquad \sigma_n(q) = 1^q + 2^q + \ldots + n^q \ (n \geq 1).$$

Then from Example 5.8, equation (5.60)

(11.8) $$\sum_{k=1}^{n}\binom{n}{k}(-1)^{k-1}\sigma_k(q) = (-1)^{n-1}(n-1)!S(q+1,n).$$

This repeats (8.23) for convenience. Notice that the terms of this sequence are zeros when $n > q+1$. See also (5.62).

Let now $F_k, L_k, k = 0,1,...,$ be the Fibonacci and Lucas numbers. Many of the following identities are from Example 5.7.

(11.9) $$\sum_{k=0}^{n}\binom{n}{k}F_k = F_{2n}$$

(11.10) $$\sum_{k=0}^{n}\binom{n}{k}F_{k+j} = F_{2n+j}, \quad j = 0,1,...$$

(11.11) $$\sum_{k=0}^{n}\binom{n}{k}kF_k = nF_{2n-1}$$

(11.12) $$\sum_{k=0}^{n}\binom{n}{k}k^2 F_k = n^2 F_{2n-2} + nF_{2n-3} \quad (n \geq 2)$$

(11.13) $$\sum_{k=0}^{n}\binom{n}{k}L_{k+j} = L_{2n+j}, \quad j = 0,1,...$$

(11.14) $$\sum_{k=0}^{n}\binom{n}{k}L_k = L_{2n}$$

(11.15) $$\sum_{k=0}^{n}\binom{n}{k}(-1)^k L_k = L_n$$

which shows that the sequence L_k is invariant under the binomial transform (1.3).

$$(11.16) \qquad \sum_{k=1}^{n} \binom{n}{k} (-1)^k k L_k = n L_{n-2} \quad (n \geq 2)$$

$$(11.17) \qquad \sum_{k=1}^{n} \binom{n}{k} (-1)^k k^2 L_k = n^2 L_{n-4} + n L_{n-3} \quad (n \geq 4)$$

$$(11.18) \qquad \sum_{k=0}^{n} \binom{n}{k} (-1)^k F_{k+j} = (-1)^{j+1} F_{n-j} \quad (0 \leq j \leq n).$$

In particular, when $j = 0$ in (11.18)

$$(11.19) \qquad \sum_{k=0}^{n} \binom{n}{k} (-1)^{k-1} F_k = F_n.$$

The summation here can be started from $k = 1$ and then from (5.6)

$$(11.20) \qquad \sum_{k=1}^{n} \binom{n}{k} (-1)^{k-1} \frac{F_k}{k} = \sum_{k=1}^{n} \frac{F_k}{k}$$

$$(11.21) \qquad \sum_{k=1}^{n} \binom{n}{k} \frac{F_k}{k} = \sum_{k=1}^{n} \frac{F_{2k}}{k}.$$

It easily follows from (11.15) that

$$(11.22) \qquad \sum_{k=1}^{n} \binom{n}{k} (-1)^k (L_k - 2) = L_n$$

and therefore,

$$(11.23a) \qquad \sum_{k=1}^{n} \binom{n}{k} (-1)^{k-1} \frac{L_k - 2}{k} = \sum_{k=1}^{n} \frac{L_k}{k}$$

or,

(11.23b)
$$\sum_{k=1}^{n}\binom{n}{k}(-1)^{k-1}\frac{L_k}{k} = 2H_n + \sum_{k=1}^{n}\frac{L_k}{k}$$

(11.24)
$$\sum_{k=0}^{n}\binom{n}{k}(-1)^{k-1}kF_k = nF_{n-2} \quad (n \geq 2)$$

(11.25)
$$\sum_{k=0}^{n}\binom{n}{k}(-1)^{k-1}k^2F_k = n^2F_{n-4} + nF_{n-3} \quad (n \geq 4)$$

(11.26)
$$\sum_{k=1}^{n}\binom{n}{k}(-1)^{k-1}\frac{F_k}{k+1} = \frac{1}{n+1}\sum_{k=1}^{n}F_k = \frac{1}{n+1}(F_{n+2}-1)$$

(11.27)
$$\sum_{k=0}^{n}\binom{n}{k}(-1)^{k}\frac{L_k}{k+1} = \frac{1}{n+1}\sum_{k=0}^{n}L_k = \frac{1}{n+1}(L_{n+2}-1).$$

For all $m \geq 0$ we have

(11.28)
$$\sum_{k=0}^{n}\binom{n}{k}(-1)^{k}\frac{L_{mk}}{L_m^k} = \frac{L_{mn}}{L_m^n}$$

(Problem B-1191 in the Fibonacci Quarterly, 55(3), 2017, p. 278.)

(11.29)
$$\sum_{k=0}^{n}\binom{n}{k}(-1)^{k}F_{2k+j} = (-1)^{n}F_{n+j} \quad (0 \leq j).$$

Now recall Binet's formulas

(11.30)
$$F_n = \frac{\varphi^n - \psi^n}{\varphi - \psi} = \frac{\varphi^n - \psi^n}{\sqrt{5}}, \quad L_n = \varphi^n + \psi^n$$

where $\varphi = \dfrac{1+\sqrt{5}}{2}$, $\psi = \dfrac{1-\sqrt{5}}{2}$. It follows that for any x and any $n \geq 0$

(11.31) $$\sum_{k=0}^{n}\binom{n}{k}x^k F_k = \frac{(1+\varphi x)^n - (1+\psi x)^n}{\sqrt{5}}$$

(11.32) $$\sum_{k=0}^{n}\binom{n}{k}x^k L_k = (1+\varphi x)^n + (1+\psi x)^n.$$

Using also the equations $\varphi + 2 = \varphi\sqrt{5}$ and $\psi + 2 = -\psi\sqrt{5}$, we find for $n > 0$

(11.33a) $$\sum_{k=0}^{n}\binom{n}{k}\varphi^k F_k = \frac{(\varphi+2)^n}{\sqrt{5}} = \left(\sqrt{5}\right)^{n-1}\varphi^n$$

(11.33b) $$\sum_{k=0}^{n}\binom{n}{k}\psi^k F_k = \frac{-(\psi+2)^n}{\sqrt{5}} = \left(-\sqrt{5}\right)^{n-1}\psi^n$$

(11.34a) $$\sum_{k=0}^{n}\binom{n}{k}\varphi^k L_k = (\varphi+2)^n = \varphi^n\left(\sqrt{5}\right)^n$$

(11.34b) $$\sum_{k=0}^{n}\binom{n}{k}\psi^k L_k = (\psi+2)^n = \left(-\psi\sqrt{5}\right)^n$$

(11.35) $$\sum_{k=0}^{n}\binom{n}{k}2^{-k} F_k = \frac{(\varphi+2)^n - (\psi+2)^n}{2^n\sqrt{5}}$$

$$= \left(\frac{\sqrt{5}}{2}\right)^n \frac{\left(\varphi^n - (-1)^n\psi^n\right)}{\sqrt{5}}$$

(see also Griffiths [38]).

Since $1+2\varphi = \varphi^3$, $1+2\psi = \psi^3$ we find from (11.31) also

(11.36a) $$\sum_{k=0}^{n}\binom{n}{k}2^k F_k = F_{3n}$$

(11.36b)
$$\sum_{k=0}^{n} \binom{n}{k} 2^k L_k = L_{3n}$$

(11.37a)
$$\sum_{k=0}^{n} \binom{n}{k} (-2)^k F_k = \left(\sqrt{5}\right)^{n-1} \left((-1)^n - 1\right)$$

(11.37b)
$$\sum_{k=0}^{n} \binom{n}{k} (-2)^k L_k = \left(\sqrt{5}\right)^{n} \left((-1)^n + 1\right).$$

The Pell numbers P_k are similar to the Fibonacci and Lucas numbers. They are defined by the recurrence

(11.38)
$$P_n = 2P_{n-1} + P_{n-2}, \quad P_0 = 0, P_1 = 1.$$

We have

(11.39)
$$\sum_{k=0}^{n} \binom{n}{k} 2^k P_k = P_{2n}.$$

For the Bernoulli numbers B_n and polynomials $B_n(x)$, for the Genocchi numbers G_n and polynomials $G_n(x)$, and also for the Euler numbers E_n and polynomials $E_n(x)$ we have

(11.40)
$$\sum_{k=0}^{n} \binom{n}{k} B_k = (-1)^n B_n$$

and in particular,

(11.41)
$$\sum_{k=0}^{n} \binom{n}{k} B_k = B_n \quad (n \neq 1)$$

(11.42)
$$\sum_{k=0}^{n} \binom{n}{k} (-1)^k B_k = (-1)^n B_n + n.$$

The last identity is proved by using Euler's transformation (3.13) as in Example 3.1. This is a short exercise for the reader.

To better understand these formulas remember that $B_n = 0$ when $n > 1$ and n is odd. Also,

$$(11.43) \qquad \sum_{k=0}^{n} \binom{n}{k} 2^k B_k(x) = 2 B_k(2x) - 2^n B_n(x)$$

$$(11.44) \qquad \sum_{k=0}^{n} \binom{n}{k} 2^k B_k = 2^n B_n\left(\frac{1}{2}\right) = 2(1 - 2^{n-1})B_n$$

$$(11.45) \qquad \sum_{k=0}^{n} \binom{n}{k} \frac{B_k}{x^k} = \frac{B_n(x)}{x^n}$$

$$(11.46) \qquad \sum_{k=0}^{n} \binom{n}{k} \frac{G_k}{x^k} = \frac{G_n(x)}{x^n}$$

$$(11.47) \qquad \sum_{k=0}^{n} \binom{n}{k} \frac{B_k}{p^k} = \frac{1}{p^n}\left(B_n + n \sum_{k=1}^{p-1} k^{n-1} \right)$$

(here $p > 0$ is an integer. This is a version of Bernoulli's formula for the sum of powers).

$$(11.48a) \qquad \sum_{k=0}^{n} \binom{n}{k} \frac{E_k}{(2x-1)^k} = \frac{2^n E_n(x)}{(2x-1)^n}$$

or in the form

$$(11.48b) \qquad \sum_{k=0}^{n} \binom{n}{k} E_k x^k = (2x)^n E_n\left(\frac{x+1}{2x}\right).$$

For $x = -1$ here we find

$$(11.49) \qquad \sum_{k=0}^{n} \binom{n}{k} (-1)^k E_k = (-2)^n E_n(0) = 2^n E_n(1)$$

and since $E_k = 0$ for k odd, we can write

(11.50)
$$\sum_{k=0}^{n}\binom{n}{k}E_k = (-2)^n E_n(0) = 2^n E_n(1).$$

Note that $E_n(0) = E_n(1) = 0$ when $n \geq 2$ is even, and $E_n(0) = -E_n(1)$ when $n \geq 1$ is odd. We also have the representation

$$B_{n+1} = -\frac{n+1}{2(2^{n+1}-1)}E_n(0) \quad (n \geq 0)$$

(11.51)
$$\sum_{k=0}^{n}\binom{n}{k}\frac{B_k(x)}{y^k} = \frac{B_n(x+y)}{y^n}$$

or using the reciprocal of y

(11.52)
$$\sum_{k=0}^{n}\binom{n}{k}y^k B_k(x) = y^n B_n(x+y^{-1}).$$

Starting the summation from $k = 1$ equation (11.51) can be written in the form

$$\sum_{k=1}^{n}\binom{n}{k}\frac{B_k(x)}{y^k} = \frac{B_n(x+y)}{y^n} + 1$$

and then from (5.6) it follows that

(11.53)
$$\sum_{k=1}^{n}\binom{n}{k}\frac{B_k(x)}{y^k k} = \sum_{k=1}^{n}\frac{B_k(x+y)}{y^k k} + H_n.$$

Similar variations can be made with many other identities. For example, the same is true for (11.55) below.

(11.54)
$$\sum_{k=0}^{n}\binom{n}{k}\frac{B_{k+1}(x)}{(k+1)y^{k+1}} = \frac{B_{n+1}(x+y)}{(n+1)y^{n+1}} - \frac{1}{n+1}$$

$$(11.55) \qquad \sum_{k=0}^{n} \binom{n}{k} \frac{E_k(x)}{y^k} = \frac{E_n(x+y)}{y^n}$$

$$(11.56) \qquad \sum_{k=0}^{n} \binom{n}{k} \frac{G_k(x)}{y^k} = \frac{G_n(x+y)}{y^n}$$

$$(11.57) \qquad \sum_{k=0}^{n} \binom{n}{k} B_k(x) = n x^{n-1} + B_n(x) = B_n(x+1)$$

$$(11.58) \qquad \sum_{k=0}^{n} \binom{n}{k} E_k(x) = 2 x^n - E_n(x) = E_n(x+1) .$$

In particular, when $x = 0$ in (11.58) and $n > 0$

$$(11.59) \qquad \sum_{k=0}^{n} \binom{n}{k} E_k(0) = E_n(1) = -E_n(0) .$$

$$(11.60) \qquad \sum_{k=0}^{n} \binom{n}{k} \frac{2}{k+1} (1 - 2^{k+1}) \frac{B_{k+1}}{x^k} = \frac{E_n(x)}{x^n}$$

$$(11.61) \qquad \sum_{k=0}^{n} \binom{n}{k} \frac{G_{k+1}}{(k+1) x^k} = \frac{E_n(x)}{x^n}$$

and by inversion

$$(11.62) \qquad \sum_{k=0}^{n} \binom{n}{k} (-1)^k \frac{E_k(x)}{x^k} = \frac{(-1)^n G_{n+1}}{(n+1) x^n}$$

$$= \frac{2(-1)^n}{n+1} (1 - 2^{n+1}) \frac{B_{n+1}}{x^n} .$$

For the Apostol-Bernoulli functions $\beta_n(x,\lambda)$ and $\beta_n(\lambda)$ we have

$$(11.63) \qquad \sum_{k=0}^{n} \binom{n}{k} \frac{\beta_k(\lambda)}{x^k} = \frac{\beta_n(x,\lambda)}{x^n}$$

$$(11.64) \qquad \sum_{k=0}^{n} \binom{n}{k} \frac{\beta_k(x,\lambda)}{y^k} = \frac{\beta_n(x+y,\lambda)}{y^n}.$$

For the exponential polynomials $\varphi_n(x)$ we have (see [5])

$$(11.65) \qquad \sum_{k=0}^{n} \binom{n}{k} \varphi_k(x) = \frac{1}{x}\varphi_{n+1}(x).$$

In particular, for the Bell numbers $\varphi_n(1)$

$$(11.66) \qquad \sum_{k=0}^{n} \binom{n}{k} \varphi_k(1) = \varphi_{n+1}(1).$$

From (11.65) by inversion

$$(11.67) \qquad \sum_{k=0}^{n} \binom{n}{k} (-1)^k \varphi_{k+1}(x) = (-1)^n x \varphi_n(x).$$

By using the property $\varphi_{n+1}(x) = x\left(\varphi_n(x) + \varphi_n'(x)\right)$ (see [5]) we find

$$(11.68) \qquad \sum_{k=0}^{n} \binom{n}{k} \varphi_k(x) = \varphi_n(x) + \varphi_n'(x).$$

For the geometric polynomials $\omega_{n,r}(x)$ with $r \geq 1$

$$(11.69) \qquad \sum_{k=0}^{n} \binom{n}{k} \omega_{k,r}(x) = \omega_{n,r}(x)\left(1 + \frac{1}{x}\right) - \frac{1}{x}\omega_{n,r-1}(x)$$

and when $r = 1$ with $\omega_k(x) = \omega_{k,1}(x)$ we have

$$(11.70) \qquad \sum_{k=0}^{n} \binom{n}{k} \omega_k(x) = \omega_n(x)\left(1 + \frac{1}{x}\right) - \frac{1}{x}\delta_{n0}$$

with the Kronecker delta. By inversion

$$(11.71) \qquad \sum_{k=0}^{n} \binom{n}{k}(-1)^k \omega_k(x) = \frac{1}{x+1}\left[(-1)^n x\omega_n(x) + 1\right].$$

For the preferential arrangement numbers $\omega_k(1)$, $k \geq 1$ we have from (11.70) and (11.71)

$$(11.72) \qquad \sum_{k=0}^{n} \binom{n}{k} \omega_k(1) = 2\omega_n(1) - \delta_{n0}$$

$$(11.73) \qquad \sum_{k=0}^{n} \binom{n}{k}(-1)^k \omega_k(1) = \frac{1}{2}\left[(-1)^n \omega_n(1) + 1\right].$$

Also

$$(11.74) \qquad \sum_{k=0}^{n} \binom{n}{k} \omega_k\left(\frac{-1}{2}\right)\frac{1}{x^k} = \frac{1}{x^n} E_n(x)$$

because of the equation

$$(11.75) \qquad \omega_k\left(\frac{-1}{2}\right) = 2B_{k+1}\frac{1 - 2^{k+1}}{k+1}$$

and (11.60). A direct proof of (11.75) is given in [2].

When $\lambda \neq 1$

$$(11.76) \qquad \sum_{k=1}^{n} \binom{n}{k}\frac{k}{x^k} \omega_{k-1}\left(\frac{\lambda}{1-\lambda}\right) = (\lambda - 1)\frac{B_n(x,\lambda)}{x^n}.$$

For the Eulerian polynomials $A_n(x)$ we have

(11.77) $$\sum_{k=0}^{n}\binom{n}{k}\frac{A_k(x)}{(1-x)^k} = \frac{A_n(x)}{x(1-x)^n}+\left(1-\frac{1}{x}\right)\delta_{n0}$$

(11.78) $$\sum_{k=1}^{n}\binom{n}{k}\frac{k}{x^k(1-\lambda)^k}A_{k-1}(\lambda) = -\frac{\beta_n(x,\lambda)}{x^n}.$$

From (7.7)

(11.79) $$\sum_{k=0}^{n}\binom{n}{k}S(k,m)x^k = \frac{1}{m!}\sum_{j=0}^{m}\binom{m}{j}(-1)^{m-j}(1+jx)^n.$$

For $x=1$ and in view of (2.1) this becomes (8.24), namely,

(11.80) $$\sum_{k=0}^{n}\binom{n}{k}S(k,m) = S(n+1,m+1).$$

Next we have

(11.81) $$\sum_{k=0}^{n}\binom{n}{k}\frac{(-x)^k}{k!} = L_n(x)$$

where

(11.82) $$L_n(x) = \frac{e^x}{n!}\left(\frac{d}{dx}\right)^n(x^n e^{-x}) \quad (n=0,1,...)$$

are the (zero order) Laguerre polynomials [58]. Dividing by x and integrating in (11.81) we obtain also

(11.83) $$\sum_{k=1}^{n}\binom{n}{k}\frac{(-x)^k}{k!k} = \int_0^x\frac{L_n(t)-1}{t}dt$$

(11.84) $$\sum_{k=0}^{n}\binom{n}{k}\lambda^k L_k(x) = (1+\lambda)^n L_n\left(\frac{x\lambda}{1+\lambda}\right).$$

Let again $\varphi = \dfrac{1+\sqrt{5}}{2}$, $\psi = \dfrac{1-\sqrt{5}}{2}$. For the Fibonacci numbers F_k, the Lucas numbers L_k, and the Laguerre polynomials $L_k(x)$ we have from (11.29) and (11.81)

(11.85)
$$\sum_{k=0}^{n} \binom{n}{k} \frac{(-1)^k F_k}{k!} = \frac{1}{\sqrt{5}}\left(L_n(\varphi) - L_n(\psi)\right)$$

(11.86)
$$\sum_{k=0}^{n} \binom{n}{k} \frac{F_k}{k!} = \frac{1}{\sqrt{5}}\left(L_n(-\varphi) - L_n(-\psi)\right)$$

(11.87)
$$\sum_{k=0}^{n} \binom{n}{k} \frac{(-1)^k L_k}{k!} = L_n(\varphi) + L_n(\psi)$$

(11.88)
$$\sum_{k=0}^{n} \binom{n}{k} \frac{L_k}{k!} = L_n(-\varphi) + L_n(-\psi).$$

Let now

(11.89)
$$H_n(x) = (-1)^n e^{x^2} \left(\frac{d}{dx}\right)^n e^{-x^2} \quad (n = 0,1,...)$$

be the Hermite polynomials defined also by the generating function (Temme [69])

(11.90)
$$e^{2xt - t^2} = \sum_{n=0}^{\infty} \frac{H_n(x)}{n!} t^n.$$

We have

(11.91)
$$\sum_{k=0}^{n} \binom{n}{k} (-1)^k \frac{H_k(x)}{(4x)^k} = \frac{H_n(x)}{(4x)^n}$$

(11.92)
$$\sum_{k=0}^{n}\binom{n}{k}\frac{H_k(x)}{(2y)^k} = \frac{H_n(x+y)}{(2y)^n}.$$

In particular, when $y = \dfrac{1}{2t}$

(11.93)
$$\sum_{k=0}^{n}\binom{n}{k}t^k H_k(x) = t^n H_n\left(x+\frac{1}{2t}\right).$$

The following formula is not a binomial transform, but is included here because of its interesting structure. This is a version of Bernoulli's formula for the sum of powers.

(11.94)
$$\sum_{k=0}^{n}\binom{n}{k}(-1)^{n-k}\frac{p^k}{k+1}B_{n-k} = \sum_{j=1}^{p}j^n.$$

Next we give some formulas involving the Cauchy numbers c_k and d_k defined in Chapter 2 (see (2.32)–(2.37)).

(11.95)
$$\sum_{k=0}^{n}\binom{n}{k}\frac{(-1)^k d_k}{k!} = \frac{d_n}{n!}$$

(11.96)
$$\sum_{k=0}^{n}\binom{n}{k}\frac{c_k}{k!} = \sum_{k=0}^{n}\frac{d_k}{k!}$$

(11.97)
$$\sum_{k=0}^{n}\binom{n}{k}H_k\frac{c_k}{k!} = H_n\sum_{k=0}^{n}\frac{d_k}{k!}-n$$

(11.98)
$$\sum_{k=0}^{n}\binom{n}{k}\frac{(-1)^k d_k}{k!}H_k = \frac{d_n}{n!}H_n-1 \quad (n\geq 1)$$

(11.99)
$$\sum_{k=0}^{n}\binom{n}{k}\frac{(-1)^k d_k}{k!}(H_k^2+H_k^{(2)})$$

$$= \frac{d_n}{n!}(H_n^2 + H_n^{(2)}) - H_n - \frac{1}{n} \quad (n \geq 1)$$

(11.100a)
$$\sum_{k=0}^{n} \binom{n}{k}(-1)^{n-k} k! = D(n)$$

or

(11.100b)
$$\sum_{k=0}^{n} \binom{n}{k}(-1)^{k} k! = (-1)^{n} D(n)$$

where $D(n)$ are the derangement numbers (the number of permutations of n elements without fixed points). By inversion in (11.100a) we find

(11.101)
$$\sum_{k=0}^{n} \binom{n}{k} D(k) = n!.$$

By a direct computation

(11.102)
$$\sum_{k=0}^{n} \binom{n}{k} k! = n!\left(1 + \frac{1}{2!} + \ldots + \frac{1}{n!}\right).$$

At the same time with $\alpha = n$ in (2.1) we have

$$1 = S(n,n) = \frac{1}{n!}\sum_{k=0}^{n} \binom{n}{k}(-1)^{n-k} k^{n}$$

that is,

(11.103)
$$\sum_{k=0}^{n} \binom{n}{k}(-1)^{n-k} k^{n} = n!.$$

This equation is not a binomial transform, as k^n depends on n. It is included here for comparison with (11.101).

Chapter 12

Transforms of Trigonometric and Hyperbolic Functions and Applications to Some Trigonometric Integrals

12.1 Binomial Transforms of Exponential, Trigonometric, and Hyperbolic Functions

(12.1)
$$\sum_{k=0}^{n}\binom{n}{k}(-1)^k e^{2ikt} = (-2i)^n e^{int}\sin^n t$$

(12.2)
$$\sum_{k=0}^{n}\binom{n}{k}(-1)^k e^{-2ikt} = (2i)^n e^{-int}\sin^n t .$$

These two identities are obtained by applying the binomial formula to the expression $(2i\sin t)^n = (e^{it} - e^{-it})^n$ written first in the form $e^{-itn}(e^{2it} - 1)^n$ and then in the form $e^{itn}(1 - e^{-2it})^n$.

In a similar manner (see Schwatt [61], pp. 214–215, and also [57])

(12.3)
$$\sum_{k=0}^{n}\binom{n}{k}\cos(kx+a) = 2^n \cos\left(\frac{nx}{2}+a\right)\left(\cos\frac{x}{2}\right)^n$$

(12.4)
$$\sum_{k=0}^{n}\binom{n}{k}\sin(kx+a) = 2^n \sin\left(\frac{nx}{2}+a\right)\left(\cos\frac{x}{2}\right)^n .$$

These two formulas were known to Ramanujan (Entry 5, p. 246 in [1]). Next,

(12.5)
$$\sum_{k=0}^{n} \binom{n}{k} (-1)^k \cos(kx + a)$$

$$= (-1)^n 2^n \cos\left(\frac{n(x+\pi)}{2} + a\right)\left(\sin\frac{x}{2}\right)^n$$

(12.6)
$$\sum_{k=0}^{n} \binom{n}{k} (-1)^k \sin(kx + a)$$

$$= (-1)^n 2^n \sin\left(\frac{n(x+\pi)}{2} + a\right)\left(\sin\frac{x}{2}\right)^n$$

(12.7)
$$\sum_{k=0}^{n} \binom{n}{k} (-1)^k \frac{\cos kx}{(\cos x)^k} = \begin{cases} 0 & (n \text{ odd}) \\ (-1)^{\frac{n}{2}} \tan^n x & (n \text{ even}) \end{cases}$$

(12.8)
$$\sum_{k=0}^{n} \binom{n}{k} (-1)^{k+1} (\cos x)^k \cos kx$$

$$= \begin{cases} (-1)^{\frac{n+1}{2}} \sin^n x \sin x & (n \text{ odd}) \\ (-1)^{\frac{n+1}{2}} \sin^n x \cos x & (n \text{ even}) \end{cases}.$$

(See also p. 119 in Loren C. Larson's book *Problem-Solving Through Problems*, Springer, 1983.)

The Chebyshev polynomials $T_n(x)$ can be defined by the equation

(12.9)
$$T_n(\cos\theta) = \cos(n\theta)$$

for $n = 0, 1, \dots$.

In terms of T_n equations (12.3) and (12.5) can be written in the form

$$(12.10) \qquad \sum_{k=0}^{n} \binom{n}{k} T_k(\cos\theta) = 2^n \cos\frac{n\theta}{2} \left(\cos\frac{\theta}{2}\right)^n$$

$$(12.11) \qquad \sum_{k=0}^{n} \binom{n}{k} (-1)^k T_k(\cos\theta) = (-1)^n 2^n \cos\frac{n(\theta+\pi)}{2} \left(\sin\frac{\theta}{2}\right)^n.$$

Similar to (12.3) and (12.4) we have for the hyperbolic functions

$$(12.12) \qquad \sum_{k=0}^{n} \binom{n}{k} \cosh(kx+a) = 2^n \cosh\left(\frac{nx}{2}+a\right) \left(\cosh\frac{x}{2}\right)^n$$

$$(12.13) \qquad \sum_{k=0}^{n} \binom{n}{k} \sinh(kx+a) = 2^n \sinh\left(\frac{nx}{2}+a\right) \left(\cosh\frac{x}{2}\right)^n$$

and similar to (12.5) and (12.6) we have also

$$(12.14) \qquad \sum_{k=0}^{n} \binom{n}{k} (-1)^k \cosh(kx+a)$$

$$= 2^{n-1} \left[1+(-1)^n\right] \cosh\left(\frac{n(x+\pi)}{2}+a\right) \left(\sin\frac{x}{2}\right)^n$$

$$-2^{n-1} \left[1+(-1)^n\right] \sinh\left(\frac{n(x+\pi)}{2}+a\right) \left(\sin\frac{x}{2}\right)^n$$

$$(12.15) \qquad \sum_{k=0}^{n} \binom{n}{k} (-1)^k \sinh(kx+a)$$

$$= 2^{n-1} \left[1+(-1)^n\right] \sinh\left(\frac{n(x+\pi)}{2}+a\right) \left(\sin\frac{x}{2}\right)^n$$

$$-2^{n-1}\left[1+(-1)^n\right]\cosh\left(\frac{n(x+\pi)}{2}+a\right)\left(\sin\frac{x}{2}\right)^n.$$

12.2 Evaluation of Integrals by Using Binomial Identities

Example 12.1 Integrating equation (12.1) with respect to t we find

(12.16) $$\int e^{int}\sin^n t\,dt = \frac{(-1)^n}{(2i)^n}\left(t+\sum_{k=1}^{n}\frac{(-1)^k e^{2ikt}}{2ik}\right)+C.$$

From here, according to (9.28)

(12.17) $$\int_{0}^{\pi/2} e^{int}\sin^n t\,dt = \frac{(-1)^n}{(2i)^n}\left(\frac{\pi}{2}+\sum_{k=1}^{n}\binom{n}{k}\frac{1}{2ik}\right)$$

$$=\frac{(-1)^n}{(2i)^n}\left(\frac{\pi}{2}+\frac{1}{2i}\left(\sum_{k=1}^{n}\frac{2^k}{k}-H_k\right)\right)$$

$$=\frac{(-1)^n}{(2i)^n}\frac{\pi}{2}+\frac{(-1)^n}{(2i)^{n+1}}\left(\sum_{k=1}^{n}\frac{2^k}{k}-H_k\right).$$

Separating real and imaginary parts in this integral we find

(12.18) $$\int_{0}^{\pi/2}\cos nt\,\sin^n t\,dt$$

$$=\frac{(-1)^n\pi}{2^n}\cos\left(\frac{n\pi}{2}\right)-\frac{(-1)^n}{2^{n+1}}\sin\left(\frac{n\pi}{2}\right)\left(\sum_{k=1}^{n}\frac{2^k}{k}-H_k\right)$$

(12.19) $$\int_{0}^{\pi/2}\sin nt\,\sin^n t\,dt$$

$$= \frac{(-1)^{n+1}\pi}{2^n}\sin\left(\frac{n\pi}{2}\right) - \frac{(-1)^{n+1}}{2^{n+1}}\cos\left(\frac{n\pi}{2}\right)\left(\sum_{k=1}^{n}\frac{2^k}{k} - H_k\right).$$

Evaluating the antiderivative (12.16) between 0 and π we get

(12.20) $$\int_0^\pi e^{int}\sin^n t\, dt = \frac{(-1)^n}{(2i)^n}\frac{\pi}{2}$$

and separating real and imaginary parts

(12.21) $$\int_0^\pi \cos nt \sin^n t\, dt = \frac{\pi}{2^n}\cos\left(\frac{n\pi}{2}\right)$$

(12.22) $$\int_0^\pi \sin nt \sin^n t\, dt = \frac{\pi}{2^n}\sin\left(\frac{n\pi}{2}\right).$$

These two integrals are entries 3.632.3 and 3.631.11 in [36].

Example 12.2 For this example we set $a = 0$ in (12.3) and (12.4) and rewrite them in the form

(12.23) $$\cos(n\theta)\cos^n\theta = \frac{1}{2^n}\sum_{k=0}^{n}\binom{n}{k}\cos(2k\theta)$$

(12.24) $$\sin(n\theta)\cos^n\theta = \frac{1}{2^n}\sum_{k=1}^{n}\binom{n}{k}\sin(2k\theta).$$

Now for any $n \geq 0$

(12.25) $$\int_0^{\pi/2}\cos(n\theta)\cos^n\theta\, d\theta = \frac{\pi}{2^{n+1}}$$

and also for $n \geq 1$

$$\int_0^{\pi/2} \sin(n\theta)\cos^n\theta\, d\theta = \frac{-1}{2^{n+1}} \sum_{k=1}^{n} \binom{n}{k} \frac{(-1)^k - 1}{k}$$

$$= \frac{1}{2^{n+1}} \left[\sum_{k=1}^{n} \binom{n}{k} \frac{(-1)^{k-1}}{k} + \sum_{k=1}^{n} \binom{n}{k} \frac{1}{k} \right].$$

According to entries (9.2) and (9.28) this equals

$$\frac{1}{2^{n+1}} \left[H_n + \sum_{k=1}^{n} \frac{2^k}{k} - H_n \right] = \frac{1}{2^{n+1}} \sum_{k=1}^{n} \frac{2^k}{k}.$$

That is, for any $n \geq 1$

$$(12.26) \qquad \int_0^{\pi/2} \sin(n\theta)\cos^n\theta\, d\theta = \frac{1}{2^{n+1}} \sum_{k=1}^{n} \frac{2^k}{k}.$$

The two integrals (12.25) and (12.26) are entries 3.631.20 and 3.631.16 in [36]. The integral (12.26) and the next one below (12.27), can be found in the works of Ramanujan — see Entry 33 on p. 290 in [1].

$$(12.27) \qquad \int_0^{\pi/2} \theta \sin(n\theta)\cos^n\theta\, d\theta = \frac{1}{2^n} \sum_{k=1}^{n} \binom{n}{k} \int_0^{\pi/2} \theta \sin(2k\theta)\, d\theta$$

$$= \frac{\pi}{2^{n+2}} \sum_{k=1}^{n} \binom{n}{k} \frac{(-1)^{k-1}}{k} = \frac{\pi}{2^{n+2}} H_n$$

after integration by parts.

In the next equation the summation starts from $k = 0$ and we have to separate the first term in the sum before integration as in the integration we divide by k

$$(12.28) \qquad \int_0^{\pi/2} \theta \cos(n\theta)\cos^n\theta\, d\theta = \frac{1}{2^n} \sum_{k=0}^{n} \binom{n}{k} \int_0^{\pi/2} \theta \cos(2k\theta)\, d\theta$$

$$= \frac{1}{2^n} \int_0^{\pi/2} \theta \, d\theta + \frac{1}{2^n} \sum_{k=1}^{n} \binom{n}{k} \int_0^{\pi/2} \theta \, d\frac{\sin(2k\theta)}{2k}$$

$$= \frac{\pi^2}{2^{n+3}} + \frac{1}{2^{n+2}} \left\{ \sum_{k=1}^{n} \binom{n}{k} \frac{(-1)^k}{k^2} - \sum_{k=1}^{n} \binom{n}{k} \frac{1}{k^2} \right\}$$

$$= \frac{\pi^2}{2^{n+3}} + \frac{1}{2^{n+2}} \left\{ -\frac{1}{2} \left(H_n^2 + H_n^{(2)} \right) - \sum_{k=1}^{n} \frac{1}{k} \sum_{j=1}^{k} \frac{2^j}{j} + \frac{1}{2} \left(H_n^2 + H_n^{(2)} \right) \right\}$$

(by using (9.31), (9.9), and (9.8)). Finally,

(12.29) $$\int_0^{\pi/2} \theta \cos(n\theta) \cos^n \theta \, d\theta = \frac{\pi^2}{2^{n+3}} - \frac{1}{2^{n+2}} \sum_{k=1}^{n} \frac{1}{k} \sum_{j=1}^{k} \frac{2^j}{j}.$$

Integrating equations (12.23) and (12.24) from 0 to π we find for any $n \geq 0$

(12.30) $$\int_0^{\pi} \cos(n\theta) \cos^n \theta = \frac{\pi}{2^n}$$

(12.31) $$\int_0^{\pi} \sin(n\theta) \cos^n \theta = 0.$$

Similar to (12.27) we have

(12.32) $$\int_0^{\pi} \theta \sin(n\theta) \cos^n \theta \, d\theta = \frac{1}{2^n} \sum_{k=1}^{n} \binom{n}{k} \int_0^{\pi} \theta \sin(2k\theta) d\theta$$

$$= \frac{-\pi}{2^{n+1}} \sum_{k=1}^{n} \binom{n}{k} \frac{1}{k} = \frac{\pi}{2^{n+1}} \left(H_n - \sum_{k=1}^{n} \frac{2^k}{k} \right)$$

by (9.28). Similar to (12.28)

$$(12.33) \qquad \int_0^\pi \theta \cos(n\theta) \cos^n \theta \, d\theta = \frac{1}{2^n} \sum_{k=0}^n \binom{n}{k} \int_0^\pi \theta \cos(2k\theta) d\theta$$

$$= \frac{1}{2^n} \int_0^\pi \theta \, d\theta + \frac{1}{2^n} \sum_{k=1}^n \binom{n}{k} \int_0^\pi \theta \, d \frac{\sin(2k\theta)}{2k}$$

$$= \frac{\pi^2}{2^{n+1}} .$$

Another proof of (12.26) by a different method was given recently by T. Amdelberhan *et al.* in the paper "The integrals of Gradshteyn and Ryzhik. Part 30; Trigonometric functions" published in SCIENTIA, Ser. A, Math. Sci. 27 (2016), 47–74.

Chapter 13

Transforms of Some Special Functions

We have included here for convenience some selected entries from Chapter 5 of Brychkov's handbook [18]. That chapter contains a number of binomial transforms of special functions and polynomials.

For the Legendre polynomials $P_n(z)$ we have

$$(13.1) \quad \sum_{k=0}^{n} \binom{n}{k} t^k P_k(z) = (1 + 2tz + t^2)^{n/2} P_n\left(\frac{1+tz}{\sqrt{1+2tz+t^2}} \right).$$

In particular, when $t = 1$ and $z = \cos 2\alpha$ we have

$$(13.2) \quad \sum_{k=0}^{n} \binom{n}{k} P_k(\cos 2\alpha) = (2\cos\alpha)^n P_n(\cos\alpha)$$

and more generally

$$(13.3) \quad \sum_{k=0}^{n} \binom{n}{k} \left(\frac{\sin\alpha}{\sin(\beta-\alpha)} \right)^k P_k(\cos\beta)$$

$$= \left(\frac{\sin\beta}{\sin(\beta-\alpha)} \right)^n P_n(\cos\alpha)$$

which turns into (13.2) for $\beta = 2\alpha$. Also,

(13.4)
$$\sum_{k=0}^{n}\binom{n}{k}(-1)^k \frac{P_k(x)}{(2x)^k} = \frac{P_n(x)}{(2x)^n}.$$

These transforms can be found also in [57].
For the Chebyshev polynomials $T_n(z)$ and $U_n(z)$ we have

(13.5)
$$\sum_{k=0}^{n}\binom{n}{k}t^k T_k(z) = (1+2tz+t^2)^{n/2} P_n\left(\frac{1+tz}{\sqrt{1+2tz+t^2}}\right)$$

(13.6)
$$\sum_{k=0}^{n}\binom{n}{k}\frac{t^k}{k+1}U_k(z)$$

$$= \frac{(1+2tz+t^2)^{n/2}}{n+1}U_n\left(\frac{1+tz}{\sqrt{1+2tz+t^2}}\right)$$

For the Laguerre polynomials $L_n^\lambda(x)$ and the digamma function $\psi(x)$

(13.7)
$$\sum_{k=0}^{n}\binom{n}{k}(-1)^k \psi(k+1)L_k^\lambda(x)$$

$$= (-1)^n\left[\psi(n+1)L_n^{\lambda-n}(x) + \sum_{k=0}^{n-1}\frac{1}{n-k}L_k^{\lambda-n}(x)\right]$$

(13.8)
$$\sum_{k=0}^{n}\binom{n}{k}(-1)^k \psi(k) = \left(2-\frac{1}{n}\right)\gamma + \left(1-\frac{1}{n}\right)\psi(n+1)$$

(13.9)
$$\sum_{k=0}^{n}\binom{n}{k}(-1)^k \psi'\left(k+\frac{1}{2}\right)$$

$$= -\frac{\pi^2}{2} + \frac{2^n(n-1)!}{(2n+1)!!}\left[\gamma + \ln 4 + \psi\left(n+\frac{1}{2}\right)\right].$$

For $n \geq 1$ we have

(13.10)
$$\sum_{k=0}^{n}\binom{n}{k}(-1)^k\{\psi(zk+1)+\gamma\}$$

$$=-\int_{0}^{\infty}(1-e^{-zt})^n\,\frac{dt}{e^t-1}$$

(13.11)
$$\sum_{k=1}^{n}\binom{n}{k}(-1)^k\left\{\frac{\psi(zk+1)+\gamma}{k}\right\}$$

$$=\int_{0}^{\infty}\frac{(1-e^{-zt})^{n+1}-1+e^{-zt}}{(e^t-1)e^{-zt}}\,dt.$$

Chapter 5 of Brychkov's book [18] (pp. 411–426) contains a number of binomial transforms involving the Hypergeometric functions. We list here several selected entries. Notations and symbols are preserved as they appear in [18]. Here $(a)_k = a(a+1)...(a+k-1)$ stands for the rising factorial.

(13.12)
$$\sum_{k=0}^{n}\binom{n}{k}(-1)^k\frac{(b-a)_k}{(b)_k}\,_1F_1\left(\begin{array}{c}a;\,z\\b+k\end{array}\right)$$

$$=\frac{(a)_n}{(b)_n}\,_1F_1\left(\begin{array}{c}a+n;\,z\\b+n\end{array}\right)$$

(13.13)
$$\sum_{k=0}^{n}\binom{n}{k}(-1)^k\frac{(d)_k}{(c)_k}\,_2F_1\left(\begin{array}{c}a,b;\,z\\c+k\end{array}\right)$$

$$=\frac{(c-d)_n}{(c)_n}\,_3F_1\left(\begin{array}{c}a,b,c-d+n\\c+n,c-d;z\end{array}\right)$$

(13.14)
$$\sum_{k=0}^{n}\binom{n}{k}(-1)^k\frac{(c-a)_k}{(c)_k}\,_2F_1\left(\begin{array}{c}a,b;\,z\\c+k\end{array}\right)$$

$$= \frac{(a)_n}{(c)_n} \, {}_2F_1 \left(\begin{matrix} a+n, \, b \\ c+n; \, z \end{matrix} \right)$$

(13.15)
$$\sum_{k=0}^{n} \binom{n}{k} \frac{(c-b)_k}{(c)_k} \left(\frac{1}{1-z} \right)^k \, {}_2F_1 \left(\begin{matrix} a, b; \, z \\ c+k \end{matrix} \right)$$

$$= \frac{1}{(1-z)^n} \, {}_2F_1 \left(\begin{matrix} a-n, \, b \\ c; \, z \end{matrix} \right)$$

(13.16)
$$\sum_{k=0}^{n} \binom{n}{k} \frac{(b)_k}{(c)_k} \, z^k \, {}_2F_1 \left(\begin{matrix} a+k, b+k; \, z \\ c+k \end{matrix} \right) = {}_2F_1 \left(\begin{matrix} a+n, \, b \\ c; \, z \end{matrix} \right)$$

(13.17)
$$\sum_{k=0}^{n} \binom{n}{k} (-1)^k \, {}_2F_1 \left(\begin{matrix} a+k, \, b \\ c; \, z \end{matrix} \right)$$

$$= \frac{(b)_n}{(c)_n} (-z)^n \, {}_2F_1 \left(\begin{matrix} a+n, b+n; \, z \\ c+n \end{matrix} \right).$$

For the Tricomi confluent hypergeometric function $\Psi(a; b; z)$

(13.18) $$\sum_{k=0}^{n} \binom{n}{k} (-1)^k \frac{(-z)^k}{(b-a)_k} \Psi \left(\begin{matrix} a; \, z \\ b+k \end{matrix} \right) = \frac{(-1)^n}{(b-a)_n} \Psi \left(\begin{matrix} a-n \\ b; \, z \end{matrix} \right)$$

(13.19) $$\sum_{k=0}^{n} \binom{n}{k} (a)_k \Psi \left(\begin{matrix} a+k; \, z \\ b+k \end{matrix} \right) = \Psi \left(\begin{matrix} a; \, z \\ b+n \end{matrix} \right)$$

(13.20) $$\sum_{k=0}^{n} \binom{n}{k} (-1)^k \Psi \left(\begin{matrix} a; \, z \\ b+k \end{matrix} \right) = (-1)^n (a)_n \Psi \left(\begin{matrix} a+n; \, z \\ b+n \end{matrix} \right)$$

(13.21) $$\sum_{k=0}^{n} \binom{n}{k} (-1)^k \Psi \left(\begin{matrix} a; \, z \\ b-k \end{matrix} \right) = (a)_n \Psi \left(\begin{matrix} a+n \\ b; \, z \end{matrix} \right).$$

Appendix A

The Stirling Transform of Sequences

Given a sequence $\{a_k\}$, its *Stirling transform* $\{b_k\}$ is the new sequence defined by

(A.1)
$$b_n = \sum_{k=0}^{n} S(n,k)\, a_k$$

with inversion

(A.2)
$$a_n = \sum_{k=0}^{n} s(n,k)\, b_k$$

where $S(n,k)$ and $s(n,k)$ are the Stirling numbers of the second kind and the first kind correspondingly. The inversion is based on the formulas

(A.3a)
$$\sum_{k=0}^{m} S(m,k)\, s(k,n) = \delta_{mn} = \begin{cases} 0 & m \neq n \\ 1 & m = n \end{cases}$$

(A.3b)
$$\sum_{k=0}^{m} s(m,k)\, S(k,n) = \delta_{mn} = \begin{cases} 0 & m \neq n \\ 1 & m = n \end{cases}.$$

The Stirling transform for sequences is related to the following Euler-Stirling transform for series.

Proposition   *Given the power series*

$$f(t) = \sum_{k=0}^{\infty} \frac{t^k}{k!} a_k$$

we have with appropriate parameters λ, μ

(A.4) $\qquad f\left(\frac{\mu}{\lambda}(e^{\lambda t} - 1)\right) = \sum_{n=0}^{\infty} \frac{t^n}{n!} \left\{\sum_{k=0}^{n} S(n,k)\lambda^{n-k}\mu^k a_k\right\}$

and also

(A.5) $\qquad f\left(\frac{\mu}{\lambda}\log(1 + \lambda t)\right) = \sum_{n=0}^{\infty} \frac{t^n}{n!} \left\{\sum_{k=0}^{n} s(n,k)\lambda^{n-k}\mu^k a_k\right\}$

(Comtet [22, p. 144]).

     Many of the following examples are from Comtet [22] and Graham *et al.* [37].

(A.6) $\qquad \sum_{k=0}^{n} s(n,k) x^k = x(x-1)...(x-n+1)$

(A.7) $\qquad \sum_{k=0}^{n} S(n,k) x(x-1)...(x-k+1) = x^n$

(A.8) $\qquad \sum_{k=0}^{n} s(n,k) = 0, \quad (n > 1)$

(A.9) $\qquad \sum_{k=0}^{n} s(n,k)(-1)^k x^k = (-1)^n x(x+1)...(x+n-1)$

(A.10) $\qquad \sum_{k=0}^{n} S(n,k)(-1)^k x(x+1)...(x+k-1) = (-1)^n x^n$

(A.11) $\qquad \sum_{k=0}^{n} s(n,k)(-1)^k = (-1)^n n!$

(A.12)
$$\sum_{k=0}^{n} S(n,k)\,k!(-1)^k = (-1)^n$$

(A.13)
$$\sum_{k=0}^{n} S(n,k)\,x^k = \varphi_n(x)$$

where $\varphi_n(x)$ are the exponential polynomials (2.22)

(A.14)
$$\sum_{k=0}^{n} S(n,k) = \varphi_n(1)$$

($\varphi_n(1)$ are the Bell numbers)

(A.15)
$$\sum_{k=0}^{n} S(n,k)\,k = \varphi_{n+1}(1) - \varphi_n(1) = \Delta\varphi_n(1)$$

(A.16)
$$\sum_{k=0}^{n} S(n,k)\,k!\,x^k = \omega_n(x)$$

($\omega_n(x)$ are the geometric polynomials (2.27)). With $x = -1$ we arrive at equation (A.12), as $\omega_n(-1) = (-1)^n$.

(A.17)
$$\sum_{k=1}^{n} S(n,k)\,(k-1)!(-1)^k = \begin{cases} 0 \ (n \neq 1) \\ -1 \ (n = 1) \end{cases}$$

(see Lemma 7.3).

(A.18) $\displaystyle\sum_{k=0}^{n} S(n,k)\,(-1)^k \frac{k!}{2^k} = \omega_n\left(\frac{-1}{2}\right) = \frac{2}{n+1}(1 - 2^{n+1})B_{n+1}$

(Graham *et al.* [37], p. 317, problem 6.76).

(A.19a) $\displaystyle\sum_{k=0}^{n} S(n,k)\,k!\,\frac{x^k}{(1-x)^{k+1}} = \sum_{k=0}^{\infty} k^n x^k \quad (|x| < 1)$

(see (2.28). The finite sum version is

$$\text{(A.19b)} \qquad \sum_{k=0}^{n} S(n,k)k!x^k \left\{ \sum_{j=0}^{m-k} \binom{k+j}{k} x^j \right\} = \sum_{k=0}^{m} k^n x^k.$$

This is obtained the same way as the sum in (5.12) from Example 5.2 by computing

$$\sum_{k=0}^{m} k^n x^k = \left(x \frac{d}{dx} \right)^n (1 + x + x^2 + \ldots + x^m)$$

$$= \sum_{k=0}^{n} S(n,k) x^k \left(\frac{d}{dx} \right)^k (1 + x + x^2 + \ldots + x^m)$$

$$= \sum_{k=0}^{n} S(n,k) x^k \left\{ \sum_{j=0}^{m-k} \frac{(j+k)!}{j!} x^j \right\} = \sum_{k=0}^{n} S(n,k) k! x^k \left\{ \sum_{j=0}^{m-k} \binom{k+j}{k} x^j \right\}.$$

The infinite version (A.19a) can be obtained from the finite version (A.19b) by setting $m \to \infty$ and using the expansion

$$\sum_{j=0}^{\infty} \binom{k+j}{k} x^j = \frac{1}{(1-x)^{k+1}} \quad (|x| < 1).$$

With $x = 1$ in (A.19b) and in view of the identity

$$\sum_{j=0}^{m-k} \binom{k+j}{k} = \binom{m+1}{k+1}$$

we find

$$\text{(A.20a)} \qquad \sum_{k=0}^{n} S(n,k) \binom{m+1}{k+1} k! = \sum_{k=0}^{m} k^n.$$

Replacing $m + 1$ by m we have from here

(A.20b)
$$\sum_{k=0}^{n} S(n,k) \frac{m(m-1)...(m-k)}{k+1} = \sum_{k=0}^{m-1} k^n .$$

It is well-known that

$$\sum_{k=0}^{m} k^n = \frac{1}{n+1} \left[B_{n+1}(m+1) - B_{n+1} \right]$$

and therefore, (A.20a) can be written in the form

(A.21)
$$\sum_{k=0}^{n} S(n,k) \binom{m+1}{k+1} k! = \frac{1}{n+1} \left[B_{n+1}(m+1) - B_{n+1} \right].$$

For $j > 0$ we have

(A.22)
$$\sum_{k=0}^{n} S(n,k) s(k+1, j) \frac{1}{k+1} = \frac{1}{n+1} \binom{n+1}{j} B_{n+1-j}$$

(see [37], p. 289, (6.99)). When $j = 1$ this becomes

(A.23)
$$\sum_{k=0}^{n} S(n,k) \frac{(-1)^k k!}{k+1} = B_n$$

and by inversion

(A.24)
$$\sum_{k=0}^{n} s(n,k) B_k = \frac{(-1)^n n!}{n+1}$$

(A.25)
$$\sum_{k=0}^{n} S(n,k) k! \frac{(-1)^k}{(k+1)^m} = (-1)^n B_n^m$$

(A.26)
$$\sum_{k=0}^{n} s(n,k) (-1)^k B_k^m = \frac{(-1)^n n!}{(n+1)^m}$$

where B_n^m are the poly-Bernoulli numbers defined by the generating function

(A.27)
$$\frac{\mathrm{Li}_m(1-e^{-x})}{1-e^{-x}} = \sum_{n=0}^{\infty} \frac{x^n}{n!} B_n^m$$

and Li_m is the polylogarithm (Kaneko [40]). Here $B_n^1 = B_n \ (n \neq 1)$.

Integrating in (A.6) with respect to x we find

(A.28)
$$\sum_{k=0}^{n} s(n,k) \frac{1}{k+1} = \int_0^1 (x)_n \, dx = c_n$$

with $(x)_n = x(x-1)...(x-n+1)$ and c_n are the Cauchy numbers of the first type (2.32). By inversion

(A.29)
$$\sum_{k=0}^{n} S(n,k) \int_0^1 (x)_n \, dx = \frac{1}{n+1}.$$

The next several formulas (and inversions) are from Wang [74], pp. 1502–1503.

(A.30)
$$\sum_{k=0}^{n} s(n,k)(-1)^k S(k+1,p+1) = (-1)^n \frac{n!}{p!} \binom{n}{p}$$

(A.31)
$$\sum_{k=0}^{n} S(n,k)(-1)^k k! \binom{k+p}{p} = (-1)^n (p+1)^n$$

(A.32)
$$\sum_{k=0}^{n} s(n,k)(-1)^k (p+1)^k = (-1)^n n! \binom{n+p}{p}$$

(A.33)
$$\sum_{k=0}^{n} S(n,k)(-1)^k k! \binom{k+p}{p} H_{k+p}$$

$$= (-1)^n n(p+1)^{n-1} + H_p(p+1)^n.$$

With $p=0$ in (A.33)

(A.34) $$\sum_{k=0}^{n} S(n,k)(-1)^k k! H_k = (-1)^n n$$

(A.35) $$\sum_{k=0}^{n} s(n,k)(-1)^k k = (-1)^n n! H_n$$

(A.36) $$\sum_{k=0}^{n} S(n,k)(-1)^k k!(H_k^2 - H_k^{(2)}) = (-1)^n n(n-1)$$

(A.37) $$\sum_{k=0}^{n} s(n,k)(-1)^k k(k-1) = (-1)^n n!(H_n^2 - H_n^{(2)})$$

(A.38) $$\sum_{k=0}^{n} S(n,k) s(k+1, p+1) = (-1)^{n-p} \binom{n}{p}$$

(when $p=0$ this turns into (A.12) since $s(k+1,1) = (-1)^k k!$). By inversion

(A.39) $$\sum_{k=0}^{n} s(n,k)(-1)^{k-p} \binom{k}{p} = s(n+1, p+1).$$

Next we list some formulas from [22] and [51] involving the Cauchy numbers c_n and d_n defined in (2.32) and (2.33):

(A.40) $$\sum_{k=0}^{n} s(n,k) \frac{1}{k+1} = c_n$$

(A.41) $$\sum_{k=0}^{n} S(n,k) c_k = \frac{1}{n+1}$$

(A.42)
$$\sum_{k=0}^{n} s(n,k)\frac{(-1)^k}{k+1} = (-1)^n d_n$$

(A.43)
$$\sum_{k=0}^{n} S(n,k)(-1)^k d_k = \frac{(-1)^n}{n+1}$$

(A.44)
$$\sum_{k=1}^{n} s(n,k)\frac{(-1)^k}{k} B_k = (-1)^{n-1}\frac{d_n}{n}$$

(A.45)
$$\sum_{k=1}^{n} S(n,k)\frac{(-1)^k d_k}{k} = -\frac{B_n}{n}.$$

The orthogonality equation (A.3b) has an important extension. For $n \ge p \ge 1$ we have (cf. (10.37))

(A.46)
$$\sum_{k=0}^{n} s(n,k)S(k,p)x^k = \frac{n!}{p!}\sum_{k=0}^{p}\binom{p}{k}(-1)^{p-k}\binom{kx}{n}.$$

For $x = 1$ this is (A.3b). For $x = -1$ we use the formula

(A.47)
$$\binom{-k}{n} = (-1)^n\binom{n+k-1}{n}$$

and (10.8) to get

(A.48)
$$\sum_{k=0}^{n} (-1)^k s(n,k)S(k,p) = (-1)^n\frac{n!}{p!}\binom{n-1}{p-1}.$$

In connection with (A.48) we note that the numbers

(A.49)
$$L(n,p) = \sum_{k=0}^{n} (-1)^{n-k} s(n,k)S(k,p) = \frac{n!}{p!}\binom{n-1}{p-1}$$

are called Lah numbers after the Slovenian mathematician and actuary Ivo Lah ([24]).

By computing $\left(x \dfrac{d}{dx} \right)^m \varphi_n(x)$ we find (see [5], Proposition 3.3])

$$(A.50) \qquad \sum_{k=0}^{n} S(n,k) k^m x^k = \sum_{j=0}^{m} \binom{m}{j} \varphi_{m-j}(-x) \varphi_{n+j}(x).$$

References

[1]   Bruce Berndt, *Ramanujans's Notebooks*, Part 1, Springer, Berlin, 1985.

[2]   Khristo N. Boyadzhiev Close encounters with the Stirling numbers of the second kind, *Math. Mag.*, 85, No. 4, October (2012), 252–266.

[3]   Khristo N. Boyadzhiev, Power sum identities with generalized Stirling numbers, *Fibonacci Quart.*, 46/47(4), November (2008/2009), 326–330.

[4]   Khristo N. Boyadzhiev, Harmonic number identities via Euler's transform, *J. Integer Seq.*, Vol. 12 (2009), Article 09.6.1.

[5]   Khristo N. Boyadzhiev, Exponential polynomials, Stirling numbers, and evaluation of some gamma integrals, *Abstr. Appl. Anal.*, Volume 2009 (2009), Article ID 168672.

[6]   Khristo N. Boyadzhiev, A series transformation formula and related polynomials, *Int. J. Math. Math. Sci.* 2005:23 (2005), 3849–3866.

[7]   Khristo N. Boyadzhiev, The Euler Series Transformation and the Binomial Identities of Ljunggren, Munarini and Simons, *Integers*, 10(3) (2010), 265–272.

[8]   Khristo N. Boyadzhiev, Series transformation formulas of Euler type, Hadamard product of functions, and harmonic number identities, *Indian J. Pure Appl. Math.*, 42 (2011), 371–387.

[9]   Khristo N. Boyadzhiev, Apostol-Bernoulli functions, derivative polynomials, and Eulerian polynomials, *Adv. Appl. Discrete Math.*, 1(2) (2008), 109–122.

[10]   Khristo N. Boyadzhiev, Binomial transform and the backward difference. *Adv. Appl. Discrete Math.*, 13(1) (2014), 43-63.

[11]   Khristo N. Boyadzhiev, Power Series with Binomial Sums and Asymptotic Expansions, *Int. J. Math. Anal.*, 8(28) (2014), 1389–1414.

[12]   Khristo N. Boyadzhiev, Series with Central Binomial Coefficients, Catalan Numbers, and Harmonic Numbers, *J. Integer Seq.* 15(2) 2012; 12.1.7.

[13]   Khristo N. Boyadzhiev, Ayhan Dil, Series with Hermite polynomials and applications, *Publ. Math. Debrecen*, 80(3–4) (2012), 387–406.

[14]   Khristo N. Boyadzhiev, Evaluation of series with binomial sums, *Anal. Math.*, 40(1) (2014), 13–23.

[15]   Khristo N. Boyadzhiev, Binomial transform of products, *Ars Comb.*, 126 (2016), 415–434.

[16]   Khristo N. Boyadzhiev, Ayhan Dil, Geometric polynomials: properties and applications to series with zeta values, *Anal. Math.*, 42(3) (2016), 203–224.

[17]   Khristo N. Boyadzhiev, Melzak's formula for arbitrary polynomials, *Util. Math*, 99 (2016), 397–401.

[18]   Yury A. Brychkov, *Handbook of Special Functions*, CRC Press, 2008.

[19]   K. W. Chen, Identities from the binomial transform, *J. Number Theory* 124 (2007), 142–150.

[20]   Wenchang Chu, Summation formulae involving harmonic numbers, *Filomat*, 26:1 (2012), 143–152.

[21]   Wenchang Chu and Qinglun Yan, Combinatorial identities on binomial coefficients and harmonic numbers, *Util. Math.*, 75 (2008), 51–66.

[22]   Louis Comtet, *Advanced Combinatorics*, Kluer, 1974.

[23]   J. L. Diaz-Barrero, J. Gibergans-Baguena, and P. G. Popescu, Some identities involving rational sums, *Appl. Anal. Discrete Math.*, 1 (2007), 397–402.

[24]   Siad Daboul *et al.*, The Lah numbers and the *n*th derivative of $e^{1/x}$, *Math Mag.*, 86(1) February (2013), 39–48.

[25]   Ayhan Dil and Veli Kurt, Polynomials related to harmonic numbers and evaluation of harmonic number series I, *Integers*, 12 (2012), A38.

[26]   Karl Dilcher, Some *q*-series identities related to divisor factors, *Discrete Math.* 145 (1995), 83–93.

[27]   Aaron Doman, An analytic proof of an interesting combinatorial identity, *Mathematical Reflections*, 4 (2014). https://www.awesomemath.org

[28]   G. P. Egorychev, Integral representation and the computation of combinatorial sums, Nauka, Novosibirsk, 1977; English translation by AMS, 1984.

[29]   Philippe Flajolet, Robert Sedgewick, Mellin transforms and asymptotics: Finite differences and Rice's integrals, *Theoret. Comput. Sci.*, 144(1–2) (1995), 101–124.

[30]   Henry W. Gould, Series transformations for finding recurrences for sequences, *Fibonacci Quart.*, 28 (1990), 166–171.

[31]   Henry W. Gould, *Combinatorial Identities*, Published by the author, Revised edition, 1972.

[32]   Henry W. Gould, Higher order extensions of Melzak's formula, *Util. Mat.* 72 (2007), 23–32.

[33]   Henry W. Gould, A note on combinatorial identities arising from the Lagrange-Waring interpolation formula, *Ars Comb.* 86 (2008), 281–288.

[34]   Henry W. Gould, Euler's formula for the n-th differences of powers, *Amer. Math. Monthly*, 85 (1978), 450–467.

[35]   Henry W. Gould, Jocelyn Quaintance, *Combinatorial identities for Stirling Numbers*, World Scientific, 2016.

[36]   Izrail S. Gradshteyn and Iosif M. Ryzhik, *Table of Integrals, Series, and Products*, Academic Press, 1980.

[37]   Ronald L. Graham, Donald E. Knuth, Oren Patashnik, *Concrete Mathematics*, Addison-Wesley Publ. Co., New York, 1994.

[38]   Martin Griffiths, Extending the domains of definition of some Fibonacci identities, *Fibonacci Quart.*, 50(4) (2012), 352–355.

[39]   Charles Jordan, *Calculus of finite differences*, Chelsea, New York, 1950 (First edition: Budapest 1939).

[40]   M. Kaneko, Poly-Bernoulli numbers, *J. Theor. Nombres de Bordeaux*, 9 (1997), 221–228.

[41]   Peter Kirschenhofer, Peter J. Larcombe, On a class of recursive-based binomial coefficient identities involving harmonic numbers, *Util. Math.* 73 (2007), 105–115.

[42]   Peter Kirschenhofer, A note on alternating sums, *Elecron. J. Combin.*, 3(2), (1996), Paper No. R7, 10pp.

[43]   K. Knopp, *Theory and Application of Infinite Series*, Dover, 1990.

[44]   Donald E. Knuth, *The Art of Computer Programming, Volume 1: Fundamental Algorithms* (3rd Edition), Addison-Wesley, 1997.

[45]   Thomas Koshy, *Fibonacci and Lucas numbers with applications*, John Wiley, New York, 2001.

[46]   Mogens E. Larsen, *Summa Summarum*, A. K. Peters, 2007.

[47]   Mogens E. Larsen and Peter J. Larcombe, Some binomial coefficient identities of specific and general type, *Util. Math.*, 74 (2007), 33–53.

[48]   L. Lewin, *Polylogarithms and associated functions*, North-Holland, Amsterdam, 1981.

[49]   Yan-Ping Mu, Symmetric recurrence relations and binomial transforms, *J. Number Theory*, 133 (2013) 3127–3137.

[50]   Z. A. Melzak, V. D. Gokhale, and W. V. Parker, Advanced Problems and Solutions: 4458, *Amer. Math. Monthly*, 60(1) (1953), 53–54.

[51]   Donatella Merlini, Renzo Sprugnoli, M. Cecilia Verri, The Cauchy numbers, *Discrete Math.*, 306 (2006), 1906–1920.

[52]   Emanuelle Munarini, Generalization of a binomial identity of Simons, *Integers*, 5 (2005), #A15.

[53]   Niels Nielsen, *Traité élémentaire des nombres de Bernoulli*, Gauthier-Villars, Paris, 1923.

[54]   N. E. Nörlund, *Vorlesungen über Differenzenrechnung*, Chelsea, 1954.

[55]   Helmut Prodinger, Some information about the binomial transform, *Fibonacci Quart.*, 32.5 (1994), 412–415.

[56]   Helmut Prodinger, Identities involving harmonic numbers that can be of interest to physicists. *Util. Math.* 83 (2010), 291–299.

[57]   A. P. Prudnikov, Yu. A. Brychkov, and O. I. Marichev, *Integrals and Series, Vol 1, Elementary Functions*, CRC, 1998.

[58]   Earl D. Rainville, *Special Functions*, Chelsea, 1971.

[59]   J. Riordan, *Combinatorial identities*, Robert E. Krieger Publ. 1979.

[60]   S. Simons, A curious identity, *The Mathematical Gazette*, 85 (2001), 296–298.

[61]   I. J. Schwatt, *An Introduction to the Operations with Series*, Chelsea, 1924.

[62]   Jonathan Sondow, Analytic continuation of Riemann's zeta function and values at negative integers via Euler's transformation of series, *Proc. Amer. Math. Soc.*, 120 (194), 412–415.

[63]   Jürgen Spieß, Some identities involving harmonic numbers, *Math. Comp.*, 55(192) (1990), 839–863.

[64]   Michael Z. Spivey, Combinatorial sums and finite differences, *Discrete Math.*, 307 (2007), 3130–3146.

[65]   Renzo Sprugnoli, *Riordan Array Proofs of Identities in Gould's Book*, 2006 (online publication).

[66] Richard P. Stanley, *Enumerative Combinatorics, Volume 1*, Cambridge University Press, Cambridge, 2012.

[67] Z. H. Sun, Invariant sequences under binomial transformation, *Fibonacci Quart.* 39(4) (2001), 324–333.

[68] Zhi-Wei Sun, Combinatorial identities in dual sequences, *European J. Combin.* 24 (2003), 709–718.

[69] Nico M. Temme, *Special Functions*, John Wiley, 1996.

[70] E. C. Titchmarsh, *The Theory of Functions*, Oxford Univ. Press, 1991.

[71] Pavel G. Todorov, Taylor expansions of analytic functions related to $(1+z)^x - 1$, *J. Math. Anal. Appl.*, 132 (1988), 264–280.

[72] Barbara Turner, Sums of powers of integers via the Binomial Theorem, *Math. Mag.*, 53(2) (1980), 92–96.

[73] Steven Vajda, *Fibonacci and Lucas Numbers, and the Golden Section: Theory and Applications*, Dover, 2007.

[74] Weiping Wang, Riordan arrays and harmonic number identities, *Comput. Math. Appl.*, 60 (2010), 1494–1509.

[75] Yi Wang, Self-inverse sequences related to a binomial inverse pair, *Fibonacci Quart.* 43(1) (2005), 46–52.

[76] Chuanan Wei (with Dian Xuan Gongand Qin Wang), Chu-Vandermonde convolution and harmonic number identities, *Integral Transforms Spec. Funct.*, 24(4) (2013), 324–330.

[77] A.N. 't Woord, Solution II to problem 10490, *Amer. Math. Monthly*, 106(6) (1999), p. 588.

[78] Aimin Xu, Zhongdi Cen, Combinatorial identities from contour integrals of rational functions, *Ramanujan J.* 40(1) (2016), 103–114.

Index